Lecture Notes in Mathematics

Edited by A. Dold and B. Eckmann

1030

T0220043

Ulrich Christian

Selberg's Zeta-, L-, and Eisensteinseries

Springer-Verlag
Berlin Heidelberg New York Tokyo 1983

Author

Ulrich Christian
Mathematisches Institut, Georg-August-Universität
Bunsenstr. 3–5, 3400 Göttingen, Federal Republic of Germany

CR Subject Classifications (1982): 3, 10

AMS Subject Classifications (1980): 10D05, 10D20, 10D24

ISBN 3-540-12701-1 Springer-Verlag Berlin Heidelberg New York Tokyo
ISBN 0-387-12701-1 Springer-Verlag New York Heidelberg Berlin Tokyo

Printing and binding: Beltz Offsetdruck, Hemsbach/Bergstr.
2146/3140-543210

PREFACE

This course of lectures was given at the University of Göttingen in the summer-semester 1983.

I thank Mrs. Christiane Gieseking for her careful typing of the troublesome manuscript.

Ulrich Christian

CONTENTS

INTRODUCTION

In these lecture notes we prove analytic continuation and functional equations for Selberg's Eisensteinseries, Selberg's zetafunctions, Selberg's L-series, and Siegel's Eisensteinseries.

We start with Epstein's zetafunction for a binary quadratic form and Epstein's L-functions which are connected with Epstein's zetafunction like Dirichlet's L-series are connected with Riemann's zetafunction. Then we consider Eisensteinseries for the elliptic modular group which are also closely related to Epstein's zetafunction.

In the next chapters we come to Selberg's zetafunction (see Maaß [33], § 17, Selberg [41], and Terras [45], [46]). Furthermore we consider Selberg's L-series which are connected with Selberg's zetafunctions like Dirichlet's L-series are connected with Riemann's zetafunction.

These functions may be described as follows. Let Y be a real symmetric, positive $n \times n$ matrix and $\overset{\nu}{Y}$ positive matrices $(\nu=1,\dots,n)$ which are connected by

(a) $$\overset{n}{Y} = Y;\quad \overset{\nu}{Y} = G'_\nu \overset{\nu+1}{Y} G_\nu \qquad (\nu = 1,\dots,n-1)$$

with integral $(\nu+1)\times\nu$ matrices G_ν. Here $'$ denotes the transposed matrix. The above mentioned authors then consider the zetafunction

(b) $$\zeta(Y;z_1,\dots,z_{n-1}) = \sum_{\langle \overset{n}{Y},\dots,\overset{1}{Y}\rangle} \prod_{\nu=1}^{n-1} (\mathrm{Det}\,\overset{\nu}{Y})^{-z_\nu}) ,$$

where the summation is taken over all possible $\overset{n-1}{Y},\dots,\overset{1}{Y}$ for which (a) holds.

Let $\Omega(n) = GL(n, \mathbb{Z})$ be the group of unimodular $n \times n$ matrices and $\Delta(n)$ the subgroup of upper triangular matrices. The above mentioned authors then show, that the function (b) is closely connected to the function

(c) $$\zeta^{*}(Y;z_1,\dots,z_{n-1}) = \sum_{U\in\Omega(n)/\Delta(n)} \Big(\prod_{\nu=1}^{n-1}(\mathrm{Det}(U'YU)_\nu)^{-z_\nu}\Big),$$

here A_ν means generally the left upper $\nu \times \nu$ submatrix of a matrix A. The summation is over all cosets $\Omega(n)/\Delta(n)$.

By computing residues Maaß [33], pages 279-299 furthermore obtaines analytic continuation of zetafunctions which are more general than (b) and (c).

In the present lecture notes we generalize the functions (b), (c) as follows. Let q be a natural number and $\chi_1,\ldots,\chi_{n-1}$ even Dirichletcharacters mod q. In (a) we assume that the elements of G_ν below the main-diagonal are divisible by q. Furthermore, we put inside the sum (b) the Dirichletcharacters $\chi_1,\ldots,\chi_{n-1}$. It is difficult to describe in the introduction how this has to be done. It is simpler for the function (c) for which we replace $\Omega(n)$ by the subgroup $\Psi(n)$ consisting of all unimodular matrices

$$U \equiv \begin{pmatrix} u_1 & * \\ 0 & u_n \end{pmatrix} \mod q .$$

Then instead of (c) we consider the function

(d) $\zeta^*(\chi_1,\ldots,\chi_{n-1};Y;z_1,\ldots,z_{n-1}) =$

$$\sum_{\Psi(n)/\Delta(n)} \prod_{\nu=1}^{n-1}(\chi_\nu(u_\nu)(\mathrm{Det}(U'YU)_\nu)^{-z_\nu}) .$$

Under the assumption that all the products $\chi_\mu \cdots \chi_\nu$ $(1 \le \mu \le \nu \le n-1)$ are primitive characters mod q we derive analytic continuation and functional equations for our functions.

For these functions we prove results that may be described as follows. Choose even Dirichletcharacters $\psi_1,\ldots,\psi_n$ with $\psi_{\nu+1}^{-1}\psi_\nu = \chi_\nu$ $(\nu = 1,\ldots,n-1)$; introduce new variables $s_1,\ldots,s_n$ by $z_\nu = s_{\nu+1} - s_\nu + \frac{1}{2}$ $(\nu = 1,\ldots,n-1)$ and put $\psi = (\psi_1,\ldots,\psi_n)$; $s = (s_1,\ldots,s_n)$. Let $L(\chi,s)$ be Dirichlet's L-series and put

$$\xi(\chi,s) = (\tfrac{\pi}{q})^{-\frac{s}{2}}\Gamma(\tfrac{s}{2})L(\chi,s).$$

If then all characters $\psi_\nu^{-1}\psi_\mu$ $(1 \le \mu < \nu \le n)$ are primitive, the function

$$\lambda(\psi,Y,s) = \Big(\prod_{1\le\mu<\nu\le n} \xi(\psi_\nu^{-1}\psi_\mu, 2(s_\nu-s_\mu)+1)\Big) \times$$

$$(\mathrm{Det}\ Y)^{s_n-\frac{1}{n}(s_1+\ldots+s_n)+\frac{n-1}{4}} \zeta^*(\chi,Y,z)$$

can be holomorphically continued to all $s \in \mathbb{C}^n$. Furthermore $\lambda(\psi,Y,s)$ satisfies certain functional equations which we shall now describe.

Let

$\tilde{s} = (-s_n,\ldots,-s_1)$; $\tilde{\psi} = (\psi_n^{-1},\ldots,\psi_1^{-1})$; $\check{s} = (-s_{n-1},\ldots,-s_1,-s_n)$;

$\check{\psi} = (\psi_{n-1}^{-1},\ldots,\psi_1^{-1},\psi_n^{-1})$; $\mathfrak{z}s = (s_n,s_1,\ldots,s_{n-1})$;

$\mathfrak{z}\psi = (\psi_n,\psi_1,\ldots,\psi_{n-1})$; $\hat{s} = (s_{n-1},s_1,\ldots,s_{n-2},s_n)$;

$\hat{\psi} = (\psi_{n-1},\psi_1,\ldots,\psi_{n-2},\psi_n)$.

Form the Gaussian sum

$$G(\chi) = q^{-\frac{1}{2}} \sum_{d \bmod q} \chi(d)\exp\left(\frac{2\pi i d}{q}\right)$$

and put

$$\eta(\psi) = \prod_{\mu=1}^{n-1} G(\chi_\mu) .$$

Form the $n{\times}n$ matrices

$$W(n) = \begin{pmatrix} 0 & & 1 \\ & \cdot^{\cdot^{\cdot}} & \\ 1 & & 0 \end{pmatrix},\quad Q(n) = \begin{pmatrix} W(n-1) & 0 \\ 0 & q \end{pmatrix},\quad P(n) = q^{\frac{1}{n}} W(n)Q^{-1}$$

and the

$$r = q^{n-2}$$

matrices

$$K_\rho = \begin{pmatrix} P^{-1}(n-1) & 0 \\ 0 & q^{1-\frac{1}{n}} \end{pmatrix} \begin{pmatrix} 1 & 0 & 0 \\ 0 & E^{(n-2)} & 0 \\ 0 & L & 1 \end{pmatrix} \qquad (\rho=1,\ldots,r)$$

where $L = (l_2,\ldots,l_{n-1})$ runs over all r residue classes mod q. Set

$$\tilde{Y} = W(n)Y^{-1}W(n), \check{Y} = (Y[Q(n)])^{-1} .$$

Then the following functional equations hold:

$$\lambda(\tilde{\psi},\tilde{Y},\tilde{s}) = \lambda(\psi,Y,s) ,$$

$$\lambda(\psi,Y,s) = \eta(\psi) q^{2(\frac{1}{n}\sum_{\nu=1}^{n} s_\nu - s_n)} \lambda(\check{\psi},\check{Y},\check{s}) ,$$

$$\lambda(\psi,Y,s) = \eta(\psi) q^{2(s_1 - \frac{1}{n}\sum_{\nu=1}^{n} s_\nu)} \lambda(\mathfrak{z}\psi, P(n)'^{-1} Y P(n)^{-1}, \mathfrak{z} s),$$

$$\lambda(\psi,Y,s) = \eta^{-1}(\psi) q^{\frac{4}{n}\sum_{\nu=1}^{n} s_\nu - 2s_{n-1} - Zs_n + 1 - \frac{n}{2}} \sum_{\rho=1}^{r} \lambda(\hat{\psi}, K_\rho' Y K_\rho, \hat{s}).$$

The transformations $\mathfrak{z}$ and $\wedge$ generate the symmetric group $\mathcal{S}_n$.

As in the case of Riemann's zetafunction and Dirichlet's L-series however the L-series have less poles than the zetafunctions. Therefore in the case $q > 1$ poles and residues of type Maaß [33], pages 279-299 do not exist. So it is impossible to get any results about more general series by computing residues. For this reason we start already with more general series and derive the analytic continuation and functional equations for them.

Let $\mathfrak{z}(n) = \{Z = Z' = X + iY,\ Y > 0\}$ Siegel's upper halfplane of degree n and $\Gamma(n) = Sp(n, \mathbb{Z})$ Siegel's modular group of degree n. Set

(e) $$M = \begin{pmatrix} A & B \\ C & D \end{pmatrix} \in \Gamma(n)$$

with $n \times n$ matrices A, B, C, D and

(f) $$M\langle Z\rangle = (AZ + B)(CZ + D)^{-1} = X_M + iY_M .$$

Set

(g) $$M\{Z\} = CZ + D .$$

Let $s = (s_1,\dots,s_n)$ a complex variablerow and $Z \in \mathfrak{z}(n)$. We consider Selberg's Eisensteinseries

(h) $\mathcal{E}(n,r,Z,s) =$

$$\sum_{M \in \Gamma_B \backslash \Gamma(n)} (\mathrm{Det}\ M\{\overline{Z}\})^{2r} (\mathrm{Det}\ Y_M)^{s_n + \frac{n}{2} + r} \prod_{\nu=1}^{n-1} (\mathrm{Det}((Y_M)_\nu))^{s_\nu - s_{\nu+1} - \frac{1}{2}} .$$

Here $\Gamma_B(n)$ is the Borel subgroup of $\Gamma(n)$ consisting of all 2n×2n matrices

(i) $$M = \begin{pmatrix} U' & SU^{-1} \\ 0 & U^{-1} \end{pmatrix}$$

with integral n×n matrices U, S. Here $S = S'$ and U is an upper triangular matrix. For the functions (h) we prove again analytic continuation and functional equations by applying a method of Diehl [11]. Since the functional equations are very similar to those of Selberg's zetafunctions I do not write them down here.

Finally consider Siegel's Eisensteinseries

(j) $$E(n,r,Z,\omega) = \sum_{M \in \Gamma_n(n) \backslash \Gamma(n)} (\mathrm{Det}\ M\{Z\})^{-2r} (\mathrm{Det}\ Y_M)^{\omega - r} .$$

Here ω is a complex variable and $\Gamma_n(n)$ the group of matrices (i) where now U is arbitrarily unimodular. We show that Siegel's Eisensteinseries may be obtained by computing residues of Selberg's Eisensteinseries. Hence the analytic continuation and the functional equations of Selberg's Eisensteinseries give us analytic continuation and a functional equation for Siegel's Eisensteinseries.

Especially we get the following results

A) It is $E(n,r,Z,\omega)$ holomorphic at $\omega = r$ for

(k) $$r = 1,2,\ [\tfrac{n-1}{2}],\ [\tfrac{n+1}{2}] ,$$

so for these values of r the Eisensteinseries $E(n,r,Z,\omega)$ has Hecke summation.

B) It is $E(n,1,Z,1) = 0$ $(n \geq 3)$.

C) Let

$$S(r) = \begin{cases} r-2 & (3 \le r < \frac{n+2}{4}) \\ [\frac{n-1}{2}]-r & (\frac{n+2}{4} \le r \le [\frac{n-3}{2}]) \end{cases} \quad . \tag{1}$$

If $3 \le r \le [\frac{n-3}{2}]$, the Eisensteinseries $E(n,r,Z,\omega)$ has at $\omega = r$ a pole of order $S(r)$ at the most.

All functions considered in this lecture play an important rôle in the theory of Siegel's modular functions but it seems to me that they are also interesting for themselves. They are eigenfunctions of invariant differential operators (see Selberg [39] till [42] and Maaß [33]) and they may be used to describe the continuous spectrum of those differential operators. Furthermore, as we have seen they may be used to get analytic continuation of Siegel's Eisenstein-series. It is an important open question if $E(n,r,Z,\omega)$ is holomorphic at $\omega = r$ also for $3 \le r \le [\frac{n-3}{2}]$.

CHAPTER I. EPSTEIN'S ZETAFUNCTIONS OF A BINARY QUADRATIC FORM

In the first chapter we consider Epstein's zetafunction for binary quadratic forms. To this zetafunction we associate L-series in the same way as Dirichlet's L-series are associated to Riemann's zetafunction. Furthermore we consider Eisensteinseries for the elliptic modular group which are closely related to Epstein's zetafunction. For all these functions we prove analytic continuation and functional equations with the aid of thetafunctions.

----

§ 1. PRELIMINARIES

§ 1 contains some preliminary definitions and results on thetafunctions for binary quadratic forms.

A matrix $K = (k_{\iota\varkappa})$ of ρ rows and σ columns is called a $\rho\times\sigma$ matrix; Rk K is the rank, K' the transposed and $\overline{K}$ the conjugate complex matrix. Occasionally we write Dg K = $= [k_1,\dots,k_{\min(\rho,\sigma)}]$ for the diagonalmatrix formed of the diagonal-elements $k_\iota = k_{\iota\iota}$ $(\iota = 1,\dots,\min(\rho,\sigma))$ from K. In case $\rho = \sigma$ let Tr K be the trace, Det K the determinante and abs K the absolute value of Det K. Let K_ν be the upper left $\nu\times\nu$ submatrix of K. With a $\rho\times\tau$ matrix L define $K[L] = L'KL$. Let 0,E be zero- and identity-matrix. The number of rows and columns will either be seen from the connection or it will be written as upper indices in brackets. A real symmetric $\rho\times\rho$ matrix Y is called positive $(Y > 0)$ respectively semipositive $(Y \geq 0)$, if the quadratic form $Y[x]$ is positive respectively nonnegative for all real columns $x \neq 0$ with ρ elements. Let $\mathcal{Y}(n)$ denote the space of all positive symmetric $n\times n$ matrices Y. Then $\mathcal{Y}(n)$ is real and has

$$(1) \qquad d(n) = \frac{n(n+1)}{2}$$

dimensions. $Y_1 > Y_2$ is defined by $Y_1 - Y_2 > 0$ and $Y_1 \geq Y_2$ by $Y_1 - Y_2 \geq 0$. A matrix is called "integral", "rational", "real" or "complex" if all elements are in $\mathbb{Z}, \mathbb{Q}, \mathbb{R}$ or $\mathbb{C}$. Brackets of type $\langle\ \rangle$ denote the greatest common devisor of integers. "exp" is the exponential function.

Let $Y \in \mathcal{Y}(n)$. By an "isotropic vector" of Y we mean a complex column w satisfiying

$$(2) \qquad Y[w] = 0 .$$

If w is an isotropc vector of Y, obviously Yw is an isotropc vector of Y^{-1}.

THEOREM 1: Let $Y \in \mathcal{Y}(n)$; $g \in \mathbb{Z}$, $g \geq 0$; u and v be two arbitrary n-rowed complex columns, w an isotropic vector of Y. Then

$$(3) \qquad \sum_{m \in \mathbb{Z}^n} ((m+v)'Yw)^g \exp(-\pi Y[m+v] + 2\pi i m'u) =$$

$$\frac{\exp(-2\pi i u'v)}{i^g} (\operatorname{Det} Y)^{-\frac{1}{2}} \sum_{m \in \mathbb{Z}^n} ((m-u)'w)^g \exp(-\pi Y^{-1}[m-u] + 2\pi i m'v).$$

Here m runs over all integral columns with n elements.

PROOF: Use Siegel [44], Page 65, formula (57).

Let $q \in \mathbb{N}$. The concept of a character mod q, a primitive character mod q and an even character mod q is defined like in Hasse [15], § 13 or Landau [22], Kapitel 22. With an even character χ mod q and $a \in \mathbb{Z}$ form the Gaussian sum

$$(4) \qquad G(\chi,a) = q^{-\frac{1}{2}} \sum_{b \bmod q} \chi(b)\exp(2\pi i\tfrac{ab}{q}) .$$

Then if χ is even, an easy computation shows

$$(5) \qquad \overline{G(\chi,a)} = G(\overline{\chi},a)$$

THEOREM 2: Let χ be a primitive character mod q. Then

$$(6) \qquad G(\chi,a) = \overline{\chi}(a)G(\chi) \qquad (a \in \mathbb{Z}).$$

Here

$$(7) \qquad G(\chi) = G(\chi,1).$$

PROOF: See Landau [22], § 126.

Let χ_1 be an even character mod q, form the row $1 = (1,1)$, let

$$(8) \qquad Y \in \mathcal{H}(2);\ t \in \mathbb{R},\ t > 0$$

and define the thetafunction

(9) $$\check{\theta}(q,1,\chi_1,Y,t) = (\mathrm{Det}\ Y)^{\frac{1}{4}}\ t^{\frac{1}{2}} \sum_{\substack{a=(a_1,a_2)' \\ a_2 \equiv 0 \bmod q}} \chi_1(a_1)\exp(-\tfrac{\pi}{q}\ Y[a]t).$$

Put

(10) $$Q(1) = \begin{pmatrix} 1 & 0 \\ 0 & q \end{pmatrix},$$

(11) $$\check{Y} = (Y[Q(1)])^{-1}.$$

Then

(12) $$\mathrm{Det}\ \check{Y} = q^{-2}\mathrm{Det}\ Y^{-1}.$$

THEOREM 3: Let χ_1 be an even primitive character mod q. Then

(13) $$\check{\theta}(q,1,\chi_1,Y,t) = G(\chi_1)\check{\theta}(q,1,\overline{\chi_1},\check{Y},t^{-1})$$

and

(14) $$\mathrm{abs}\ G(\chi_1) = 1.$$

PROOF: In (9) set $a = b+qc$ with $b = (b_1 0)'$ and c an integral column. Then (9) gives

(15) $$\check{\theta}(q,1,\chi_1,Y,t) = (\mathrm{Det}(tY))^{\frac{1}{4}} \sum_{b_1 \bmod q} \chi_1(b_1) \sum_{c \in \mathbb{Z}^2} \exp(-\pi(qYt)[c+\tfrac{b}{q}]).$$

Apply (3) with $n=2$, $g=0$, $u=0$, $v=\frac{b}{q}$, qYt instead of Y. Then

(16) $$\check{\theta}(q,1,\chi_1,Y,t) = (\mathrm{Det}(tY))^{\frac{1}{4}}(\mathrm{Det}(qYt))^{-\frac{1}{2}} \times \sum_{c \in \mathbb{Z}^2} \Big(\sum_{b_1 \bmod q} \chi_1(b_1)\exp(2\pi i \frac{c_1 b_1}{q})\Big)\exp(-\frac{\pi}{q} Y^{-1}[c]t^{-1}).$$

From (4), (6), (12), (16) we get

(17) $$\check{\theta}(q,1,\chi_1,Y,t) = G(\chi_1)(\mathrm{Det}\ \check{Y})^{\frac{1}{4}}\ t^{-\frac{1}{2}} \sum_{c \in \mathbb{Z}^2} \bar{\chi}(c_1)\exp(-\frac{\pi}{q}\check{Y}[Q(1)c]t^{-1}) = G(\chi_1)\check{\theta}(q,1,\bar{\chi}_1,\check{Y},t^{-1}) .$$

Herewith one has formula (13). Inserting $\bar{\chi}_1$ $,\check{Y},t^{-1}$ in the left hand side of (13) and applying (13) once more one gets

(18) $$G(\chi_1)G(\bar{\chi}_1) = 1 .$$

From (5), (7), (18) we deduce (14). Theorem 3 is proved.

According to Maaß [33], pages 210, 267 form the differential operator

(19) $$D^*(t) = t^{\frac{1}{2}}\ \frac{d}{dt}\ t^2\ \frac{d}{dt}\ t^{-\frac{1}{2}} = -\ \frac{1}{4} + t\ \frac{d}{dt} + t^2\ \frac{d^2}{dt^2} .$$

Using

(20) $$\frac{d}{dt} = -t^{-2}\ \frac{d}{dt^{-1}}$$

one see's

(21) $$D^*(t) = D^*(t^{-1}),$$

Put

(22) $$D(q,t) = \begin{cases} -D^*(t) & (q = 1) \\ 1 & (q > 1 \end{cases} .$$

Then also

$$D(q,t) = D(q,t^{-1}) \ . \tag{23}$$

Put

$$\dot{\theta}(q,1,\chi_1,Y,t) = D(q,t)\check{\theta}(q,1,\chi_1,Y,t) \ . \tag{24}$$

THEOREM 4: Let χ_1 be an even primitive character mod q. Then

$$\theta(q,1,\chi_1,Y,t) = G(\chi_1)\theta(q,1,\bar{\chi}_1,\check{Y},t^{-1}) \ . \tag{25}$$

PROOF: Apply Theorem 3 and formula (23).

From (9), (19) one obtains

$$\theta(1,1,1,Y,t) = (\text{Det } Y)^{\frac{1}{4}} t^{\frac{1}{2}} \sum_{\substack{a \in \mathbb{Z}^2 \\ a \neq 0}} \{(2\pi Y[a]t)-(\pi Y[a]t)^2\} \exp(-\pi Y[a]t). \tag{26}$$

THEOREM 5: For all even characters $\chi_1 \bmod q$ one has

$$\text{abs } \theta(q,1,\chi_1,Y,t) \leq c_1(\text{Det } Y)^{\frac{1}{4}} t^{\frac{1}{2}} \sum_{\substack{a \in \mathbb{Z}^2 \\ a \neq 0}} \exp(-\tfrac{\pi}{2q} Y[a]t), \tag{27}$$

with a constant $c_1 > 1$.

PROOF: For $q = 1$, $\chi_1 = 1$ this follows from (26). For $q > 1$ it follows from (9), (22) and $\chi_1(0) = 0$.

THEOREM 6: Let R_2 be a non-singular rational 2x2 matrix, χ_1 an arbitrary even character mod q and $j(Y)$ a positive number with

(28) $$Y \geq j(Y)E .$$

Then there exists a real number $c_2 = c_2(R_2) \geq 1$ with

(29) $$\operatorname{abs} \theta(q,1,\chi_1,Y[R_2],t) \leq c_2 j(Y)^{-1}(\operatorname{Det} Y)^{\frac{1}{4}} t^{-\frac{1}{2}} \exp(-c_2^{-1} j(Y)t).$$

PROOF: Let $R_2 = rG$ with $r \in Q$, $r \neq 0$ and G integral. If a is an integral column, so is $b = Ga$ and if $a \neq 0$ then $b \neq 0$. Because of (27), (28) it suffices to prove for $u > 0$:

(30) $$\sum_{\substack{b \in \mathbb{Z}^2 \\ b \neq 0}} \exp(-2\pi u b'b) \leq d_1 u^{-1} \exp(-d_1^{-1} u)$$

with a constant $d_1 \geq 1$.

Set

(31) $$\vartheta(u) = \sum_{v=-\infty}^{\infty} \exp(-\pi v^2 u) .$$

Then from (3) we obtain

(32) $$\vartheta(u) = u^{-\frac{1}{2}} \vartheta(u^{-1}) .$$

Furthermore

(33) $$\vartheta(u) \leq \vartheta(1) \qquad (u \geq 1).$$

Hence from (32), (33)

(34) $$\vartheta(u) \leq u^{-\frac{1}{2}} \vartheta(1) \qquad (u \leq 1).$$

From (33), (34) we get for $\epsilon > 0$ and some constant $d_2 = d_2(\epsilon) \geq 1$

(35) $$\vartheta(u) \leq d_2 u^{-\frac{1}{2}} \exp(\epsilon u) \qquad (u > 0).$$

On the left hand side of (30) is $b'b \geq 1$. Hence from (31) and (35) with $\epsilon = 1$ we get

$$\sum_{\substack{b \in \mathbb{Z}^2 \\ b \neq 0}} \exp(-2\pi u b'b) \leq \exp(-\pi u)\vartheta^2(u) \leq d_2 u^{-1} \exp(-(\pi-2)u).$$

Now (30) follows.

Let

(36) $$\mathfrak{Z}(1) = \{z = x + iy \in \mathbb{C};\ x,y \in \mathbb{R};\ y > 0\}$$

be the upper halfplane, and

(37) $$Sp(1, \mathbb{R}) = \{M = \begin{pmatrix} a & b \\ c & d \end{pmatrix};\ a,b,c,d \in \mathbb{R};\ ad - bc = 1\}$$

the symplectic group of degree one. $Sp(1, \mathbb{R})$ operates on $\mathfrak{Z}(1)$ by $z \to M\langle z\rangle$ with

(38) $$M\langle z\rangle = \frac{az + b}{cz + d} .$$

The subgroup $\Gamma(1) = Sp(1, \mathbb{Z})$ of $Sp(1, \mathbb{R})$ with integral M is the elliptic modular group. It operates discontinuously on $\mathfrak{Z}(1)$. A fundamental domain of $Sp(1, \mathbb{Z})$ is given by

(39) $$\mathfrak{F}(1) = \{z \in \mathfrak{Z}(1);\ \text{abs}\ z \geq 1;\ \text{abs}\ x \leq \tfrac{1}{2}\}.$$

Obviously

(40) $$y \geq \frac{\sqrt{3}}{2} \qquad (z \in \mathfrak{F}(1)).$$

Let $z \in \mathfrak{Z}(1)$ and set

(41) $$Y = \frac{1}{y}\begin{pmatrix} |z|^2 & x \\ x & 1 \end{pmatrix} \in \mathfrak{Y}(2).$$

Then

(42) $$\text{Det}\ Y = 1 .$$

Obviously

$$(43)\qquad Y^{-1} = Y[I] \ ,$$

with

$$(44)\qquad I = \begin{pmatrix} 0 & 1 \\ -1 & 0 \end{pmatrix} .$$

The column

$$(45)\qquad w = (i \ -i\,\bar{z})'$$

is an isotropic vector of Y. Because of (43) the column $I^{-1}w$ is an isotropic vector of Y^{-1}. Let $h = (h_1, h_2)'$ be an integral column. Then

$$(46)\qquad Y[h] = \frac{1}{y}|h_1\bar{z} + h_2|^2$$

and

$$(47)\qquad w'Yh = h_1\bar{z} + h_2 \ .$$

Let $r \in \mathbb{N}$, $t \in \mathbb{R}$, $t > 0$ and set

$$(48)\qquad \theta^*(r,z,t) = \sum_{h \in \mathbb{Z}^2} (h_1\bar{z} + h_2)^{2r} \exp(-\tfrac{\pi}{y}|h_1 z + h_2|^2 t) \ .$$

Then from (46), (47), (48) we get

$$(49)\qquad \theta^*(r,z,t) = t^{-2r} \sum_{h \in \mathbb{Z}^2} (h'tYw)^{2r} \exp(-\pi(tY)[h]).$$

The application of (3) to (49) gives

$$(50)\qquad \theta^*(r,z,t) = (-1)^r t^{-1-2r} \sum_{h \in \mathbb{Z}^2} (h'w)^{2r} \exp(-\pi Y^{-1}[h]t^{-1}).$$

Using (43) and writing h instead of Ih we get

$$(51)\qquad \theta^*(r,z,t) = t^{-1-2r} \sum_{h\in\mathbb{Z}^2} (-1)^r (h'Iw)^{2r} \exp(-\pi Y[h]t^{-1}) .$$

An easy computation shows

$$(52)\qquad (-1)^r (h'Iw)^{2r} = (h_1\bar{z} + h_2)^{2r} .$$

Inserting this in (51) and using (46), (48) we get

$$(53)\qquad \theta^*(r,z,t) = t^{-1-2r}\theta^*(r,z,t^{-1}).$$

An easy computation shows

$$(54)\qquad \theta^*(r,M\langle z\rangle,t) = (c\bar{z} + d)^{-2r}\theta^*(r,z,t) \qquad (M \in \Gamma(1)).$$

From $r > 0$ and (48) we deduce

$$(55)\qquad \operatorname{abs} \theta^*(r,z,t) \le \left(\frac{y}{t}\right)^r \sum_{\substack{h\in\mathbb{Z}^2\\ h \neq 0}} \left(\frac{t}{y}|h_1z + h_2|^2\right)^r \exp\left(-\pi\frac{t}{y}|h_1z + h_2|^2\right),$$

hence

$$(56)\qquad \operatorname{abs} \theta^*(r,z,t) \le c_3\left(\frac{y}{t}\right)^r \sum_{\substack{h\in\mathbb{Z}^2\\ h \neq 0}} \exp\left(-(\pi-\varepsilon)\frac{t}{y}|h_1z + h_2|^2\right),$$

with $0 < \varepsilon < \pi$ and some constant $c_3(\varepsilon) \ge 1$.

THEOREM 7: Let $M \in \Gamma(1)$, $z \in \mathfrak{F}(1)$. Then there exists a constant $c_4 = c_4(M) \ge 1$ with

$$(57)\qquad \operatorname{abs} \theta^*(r,M\langle z\rangle,t) \le c_4\left(\frac{y}{t}\right)^{r+1} \exp\left(-c_4^{-1}\,\frac{t}{y}\right) .$$

PROOF: For $z \in \mathfrak{F}(1)$, $M \in \Gamma(1)$ we have $|cz + d| \ge 1$. Hence by (54)

$$(58)\qquad \operatorname{abs} \theta^*(r,M\langle z\rangle,t) \le \operatorname{abs} \theta^*(r,z,t) \qquad (M \in \Gamma(1)).$$

For $z \in \mathcal{f}(1)$ we have

$$|h_1 z + h_2|^2 = |z|^2 h_1^2 + 2xh_1h_2 + h_2^2 \geq$$

$$h_1^2 - |h_1||h_2| + h_2^2 = \frac{1}{2}\, h'h + \frac{1}{2}(|h_1|-|h_2|)^2 \geq \frac{1}{2}\, h'h.$$

Hence

$$|h_1 z + h_2|^2 \geq \frac{1}{2}\, h'h \qquad (z \in \mathcal{f}(1)) .$$

Using (56) with $\epsilon = \frac{\pi}{2}$ we get

$$\text{abs}\ \theta^*(r,z,t) \leq c_3\left(\frac{y}{t}\right)^r \sum_{\substack{h \in \mathbb{Z}^2 \\ h \neq 0}} \exp\left(-2\pi\frac{t}{8y}\, h'h\right).$$

Hence from (30), (58) we get (57). Theorem 7 is proved.

§ 2. EPSTEIN'S ZETAFUNCTIONS AND L-SERIES

§ 2 contains the definition of Epstein's zetafunctions and L-series. The functions are analytically continued and functional equations are proved.

Let $\Omega(2)$ be the group of unimodular 2x2 matrices, i. e., $\Omega(2)$ consists of all integral 2x2 matrices U with abs U = 1. For $q \in \mathbb{N}$ form the subgroup

(59) $$\Psi = \{U = \begin{pmatrix} a & b \\ c & d \end{pmatrix} \in \Omega(2);\ c \equiv 0 \bmod q\}$$

and set

(60) $$\Delta = \{U = \begin{pmatrix} \pm 1 & * \\ 0 & \pm 1 \end{pmatrix} \in \Omega(2)\}$$

Let $l = (1,1)$, χ_1 an even character mod 1, $Y \in \mathcal{P}(2)$ and ω a complex variable. Define

$$(61)\qquad \hat{\zeta}(q,1,\chi_1,Y,\omega) = \frac{1}{2} \sum_{\substack{h=(h_1,h_2)'\in\mathbb{Z}^2 \\ h_2 \equiv 0 \bmod q \\ h \neq 0}} \chi_1(h_1)(Y[h])^{-\omega} ,$$

$$(62)\qquad \hat{\zeta}^*(q,1,\chi_1,Y,\omega) = \sum_{U = \binom{a\ b}{c\ d} \in \Psi/\Delta} \chi_1(a)((Y[U])_1)^{-\omega}$$

Here $(Y[U])_1$ means the upper left element of $Y[U]$ as was already defined at the beginning of § 1. It follows from Siegel [44], Chapter I, § 5 that the series (61), (62) converge absolutely for

$$(63)\qquad \operatorname{Re} \omega > 1 .$$

If $q = 1$ (and hence χ is the principal character) (61) is called "Epstein's zetafunction". For $q = 1$ we shall also call (62) an Epstein's zetafunction. If $q > 1$ we shall call (61), (62) "Epstein's L-series".

The series (62) may be written as

$$(64)\qquad \hat{\zeta}^*(q,1,\chi_1,Y,\omega) = \frac{1}{2} \sum_{\substack{\langle a,c\rangle=1 \\ c \equiv 0 \bmod q}} \chi_1(a)(Y[\binom{a}{c}])^{-\omega} .$$

In (61) we have $\chi_1(h_1) = 0$ for $\langle h_1,q\rangle > 1$. Hence it suffices to sum over those h for which

$$(65)\qquad \langle h_1,q\rangle = 1 .$$

Then one may set

$$(66)\qquad h = k\binom{a}{c} , \quad \langle a,c\rangle = 1.$$

From (65) we conclude $\langle ak,q\rangle = 1$, hence from $h_2 \equiv 0 \bmod q$ it follows that $c \equiv 0 \bmod q$. Therefore

$$(67)\qquad \hat{\zeta}(q,1,\chi_1,Y,\omega) = L(\chi_1,2\omega)\hat{\zeta}^*(q,1,\chi_1,Y,\omega),$$

where $L(\chi_1,\ldots)$ denotes Dirichlet's **L-series**.

With two complex variables s_1, s_2 set

(68) $$\omega = s_2 - s_1 + \frac{1}{2},$$

(69) $$\zeta(q,1,\chi_1,Y,s) = \hat{\zeta}(q,1,\chi_1,Y,\omega); \zeta^*(q,1,\chi_1,Y,s) = \\ = \hat{\zeta}^*(q,1,\chi_1,Y,\omega) .$$

Then

(70) $$\zeta(q,1,\chi_1,Y,s) = \frac{1}{2} \sum_{\substack{h = (h_1,h_2)' \in \mathbb{Z}^2 \\ h_2 \equiv 0 \bmod q \\ h \neq 0}} \chi_1(h_1)(Y[h])^{s_1-s_2-\frac{1}{2}},$$

(71) $$\zeta^*(q,1,\chi_1,Y,s) = \sum_{U = \begin{pmatrix} a & b \\ c & d \end{pmatrix} \in \Psi/\Delta} \chi_1(a)((Y[U])_1)^{s_1-s_2-\frac{1}{2}}.$$

If one sets $\sigma_\nu = \mathrm{Re}\, s_\nu$ $(\nu = 1,2)$, the series (70), (71) converge absolutely for

(72) $$\sigma_2 - \sigma_1 > \frac{1}{2} .$$

From (67) we get

(73) $$\zeta(q,1,\chi_1,Y,s) = L(\chi_1, 2s_2 - 2s_1 + 1)\zeta^*(q,1,\chi,Y,s)$$

THEOREM 8: Let

(74) $$U = \begin{pmatrix} u_1 & u_2 \\ u_3 & u_4 \end{pmatrix} \in \Psi .$$

Then

$$(75)\qquad \zeta^*(q,\mathfrak{l},\chi_1,Y[U],s) = \chi_1^{-1}(u_1)\zeta^*(q,\mathfrak{l},\chi_1,Y,s),$$

$$(76)\qquad \zeta(q,\mathfrak{l},\chi_1,Y[U],s) = \chi_1^{-1}(u_1)\zeta(q,\mathfrak{l},\chi_1,Y,s) .$$

PROOF: From (71) we deduce

$$(77)\qquad \zeta^*(q,\mathfrak{l},\chi_1,Y[U],s) = \sum_{V = \begin{pmatrix} a & b \\ c & d \end{pmatrix}} \chi_1(a)((Y[UV])_1)^{s_1-s_2-\frac{1}{2}} = \chi_1^{-1}(u_1) \sum_{V \in \Psi/\Delta} \chi_1(u_1 a)((Y[UV])_1)^{s_1-s_2-\frac{1}{2}} .$$

With V also UV runs over Ψ/Δ and $(UV)_1 \equiv u_1 a \bmod q$. Hence (75) follows from (77). From (73) and (75) we get (76). Theorem 8 is proved.

Set $\mathfrak{l} = (1,1)$ and

$$(78)\qquad f(\mathfrak{l},Y,s) = (\operatorname{Det} Y)^{s_2+\frac{1}{4}}(Y_1)^{s_1-s_2-\frac{1}{2}} ,$$

$$(79)\qquad W = \begin{pmatrix} 0 & 1 \\ 1 & 0 \end{pmatrix} ,$$

$$(80)\qquad \tilde{Y} = (Y[W])^{-1} = Y^{-1}[W],$$

$$(81)\qquad \tilde{s}_1 = -s_2;\ \tilde{s}_2 = -s_1;\ \tilde{s} = (\tilde{s}_1,\tilde{s}_2) .$$

THEOREM 9: The function $f(\mathfrak{l},Y,s)$ is homogeneous in Y of degree $s_1 + s_2$ and

$$(82)\qquad f(\mathfrak{l},\tilde{Y},\tilde{s}) = f(\mathfrak{l},Y,s) .$$

Let $e = (1,1)$ and $a \in \mathbb{C}$. Then

$$(83)\qquad (\operatorname{Det} Y)^a f(\mathfrak{l},Y,s) = f(\mathfrak{l},Y,ae + s) .$$

Let $D = \begin{pmatrix} d_1 & * \\ 0 & d_2 \end{pmatrix}$ be a real upper tringular matrix. Then

(84) $$f(1,Y[D],s) = f(1,Y,s)(\text{abs } d_1)^{2s_1-\frac{1}{2}}(\text{abs } d_2)^{2s_2+\frac{1}{2}} .$$

Especially

(85) $$f(1,Y[V],s) = f(1,Y,s) \qquad (V \in \Delta) .$$

PROOF: It suffices to prove (82), the rest is trivial.

Set

(86) $$Y = \begin{pmatrix} y_1 & y_{12} \\ y_{12} & y_2 \end{pmatrix} .$$

Then

(87) $$f(1,Y,s) = (\text{Det } Y)^{s_2+\frac{1}{4}} y_1^{s_1-s_2-\frac{1}{2}} ,$$

(88) $$\tilde{Y} = \begin{pmatrix} \frac{y_1}{\text{Det } Y} & -\frac{y_{12}}{\text{Det } Y} \\ -\frac{y_{12}}{\text{Det } Y} & \frac{y_2}{\text{Det } Y} \end{pmatrix} .$$

Hence

$$f(1,\tilde{Y},\tilde{s}) = ((\text{Det } Y)^{-1})^{-s_1+\frac{1}{4}} \left(\frac{y_1}{\text{Det } Y}\right)^{-s_2+s_1-\frac{1}{2}} = (\text{Det } Y)^{s_2+\frac{1}{4}} y_1^{s_1-s_2-\frac{1}{2}} =$$

$$= f(1,Y,s) .$$

This proves (82). Theorem 9 is proved.

For $U \in \Psi$ set

(89) $$\tilde{U} = WU'^{-1}W \in \Psi .$$

If U runs over Ψ/Δ also $\tilde{U}$ does. Put

$$(90) \qquad U = \begin{pmatrix} u_1 & u_2 \\ u_3 & u_4 \end{pmatrix}, \quad \tilde{U} = \begin{pmatrix} \tilde{u}_1 & \tilde{u}_2 \\ \tilde{u}_3 & \tilde{u}_4 \end{pmatrix},$$

then

$$(91) \qquad u_1\tilde{u}_4 \equiv u_4\tilde{u}_1 \equiv 1 \bmod q .$$

Set

$$(92) \qquad \Lambda(q,l,\chi_1,Y,s) = (\operatorname{Det} Y)^{s_2+\frac{1}{4}} \zeta^*(q,l,\chi_1,Y,s).$$

Then

$$(93) \qquad \Lambda(q,l,\chi_1,Y,s) = \sum_{U = \left(\begin{smallmatrix} a & b \\ c & d \end{smallmatrix}\right) \in \Psi/\Delta} \chi_1(a) f(l,Y[U],s).$$

From (75), (92) we deduce

$$(94) \qquad \Lambda(q,l,\chi_1,Y[U],s) = \chi_1^{-1}(u_1)\Lambda(q,l,\chi_1,Y,s) \qquad (U \in \Psi).$$

THEOREM 10:

$$(95) \qquad \Lambda(q,l,\chi_1,\tilde{Y},\tilde{s}) = \Lambda(q,l,\chi_1,Y,s) .$$

PROOF: From $u_1u_4 \equiv \pm 1 \bmod q$ and (91) we get

$$(96) \qquad \tilde{u}_1 \equiv \pm u_1 \bmod q, \; \tilde{u}_4 \equiv \pm u_4 \bmod q .$$

Use

$$(97) \qquad \widetilde{Y[U]} = \tilde{Y}[\tilde{U}] .$$

Then because of (82), (96), (97)

$$\Lambda(q,1,\chi_1,\tilde{Y},\tilde{s}) = \sum_{\tilde{U} \in \Psi/\Delta}' \chi_1(\tilde{u}_1) f(1,\widetilde{Y[U]},\tilde{s}) =$$

$$\sum_{U \in \Psi/\Delta}' \chi_1(u_1) f(1,Y[U],s) = \Lambda(q,1,\chi_1,Y,s).$$

Theorem 10 is proved.

THEOREM 11: Set

$$(98) \qquad \delta(q,\omega) = \begin{Bmatrix} \omega(1-\omega) & (q=1) \\ 1 & (q>1) \end{Bmatrix}.$$

Then the function

$$(99) \quad \hat{\lambda}(q,1,\chi_1,Y,\omega) = \delta(q,\omega)\left(\frac{\Pi}{q}\right)^{-\omega} \Gamma(\omega)(\mathrm{Det}\ Y)^{\frac{\omega}{2}} \hat{\zeta}(q,1,\chi_1,Y,\omega)$$

is homogeneous in Y of degree 0 and

$$(100) \quad \hat{\lambda}(q,1,\chi_1,Y,\omega) = (\mathrm{Det}\ Y)^{\frac{\omega}{2}-\frac{1}{4}} \frac{1}{2}\int_0^\infty \theta(q,1,\chi_1,Y,t)t^{\omega-\frac{1}{2}} \frac{dt}{t}.$$

Furthermore

$$(101) \qquad \delta(q,1-\omega) = \delta(q,\omega).$$

PROOF: Let $q = 1$, $\chi_1 = 1$. Then from (26) we deduce

$$(\mathrm{Det}\ Y)^{\frac{\omega}{2}-\frac{1}{4}} \frac{1}{2}\int_0^\infty \theta(1,1,1,Y,t)t^{\omega-\frac{1}{2}} \frac{dt}{t} =$$

$$(\mathrm{Det}\ Y)^{\frac{\omega}{2}} \frac{1}{2} \sum_{\substack{a \in \mathbb{Z}^2 \\ a \neq 0}} \Big\{ (2\pi Y[a]) \int_0^\infty t^{\omega+1} \exp(-\pi Y[a]t)\frac{dt}{t}$$

$$- (\pi Y[a])^2 \int_0^\infty t^{\omega+2} \exp(-\pi Y[a]t)\frac{dt}{t} \Big\} =$$

$$(\mathrm{Det}\,Y)^{\frac{\omega}{2}}\,\pi^{-\omega}\,\frac{1}{2}\sum_{\substack{a\in\mathbb{Z}^2\\ a\neq 0}}(Y[a])^{-\omega})(2\Gamma(\omega+1)-\Gamma(\omega+2))$$

$$=\ (\mathrm{Det}\,Y)^{\frac{\omega}{2}}\,\pi^{-\omega}\,\delta(1,\omega)\Gamma(\omega)\hat{\zeta}(1,1,1,Y,\omega)\ =\ \hat{\lambda}(1,1,1,Y,\omega)\ .$$

Hence for $q = 1$ the theorem is proved.

Let $q > 1$. Then from (9), (24) we get

$$(\mathrm{Det}\,Y)^{\frac{\omega}{2}-\frac{1}{4}}\,\frac{1}{2}\int_0^\infty \theta(q,1,\chi_1,Y,t)t^{\omega-\frac{1}{2}}\frac{dt}{t}\ =$$

$$(\mathrm{Det}\,Y)^{\frac{\omega}{2}}\sum_{\substack{a=(a_1a_2)'\\ a_2\equiv 0 \bmod q\\ a\neq 0}}\chi\,(a_1)\int_0^\infty t^{\omega}\exp(-\,\frac{\pi}{q}\,Y[a]t)\frac{dt}{t}$$

$$=\ (\mathrm{Det}\,Y)^{\frac{\omega}{2}}\,(\frac{\pi}{q})^{-\omega}\Gamma(\omega)\hat{\zeta}(q,1,\chi_1,Y,\omega)\ =\ \hat{\lambda}(q,1,\chi_1,Y,\omega).$$

This proves theorem 11.

We introduce the functions

(102) $\qquad \xi(\chi_1,\omega)\ =\ (\frac{\pi}{q})^{-\omega}\Gamma(\omega)L(\chi_1,2\omega),$

and for $q = 1$:

(103) $\qquad F(\omega)\ =\ \delta(1,\omega)\xi(1,\omega)\ =\ \omega(1-\omega)\pi^{-\omega}\,\Gamma(\omega)\zeta(2\omega)\ ,$

where $L(1,\omega) = \zeta(\omega)$ is Riemann's zeta function

(104) $\qquad F^*(q,\chi_1,\omega)\ =\ \delta(q,\omega)\xi(\chi_1,\omega)\ =\ \left\{\begin{matrix} F(\omega) & (q = 1) \\ \xi(\chi_1,\omega) & (q > 1)\end{matrix}\right\}\ .$

From (67), (99), (102), (103), (104) we get

$$(105)\qquad \hat{\lambda}(q,l,\chi_1,Y,\omega) = F^*(q,\chi_1,\omega)(\mathrm{Det}\ Y)^{\frac{\omega}{2}}\ \hat{\zeta}*(q,l,\chi_1,Y,\omega)\ .$$

Put

$$(106)\qquad \lambda(q,l,\chi_1,Y,s) = \hat{\lambda}(q,l,\chi_1,Y,\omega)\ .$$

Then

$$(107)\qquad \lambda(q,l,\chi_1,Y,s) = \delta(q,s_2-s_1+\tfrac{1}{2})(\tfrac{\Pi}{q})^{s_1-s_2-\frac{1}{2}}\Gamma(s_2-s_1+\tfrac{1}{2})\times$$

$$(\mathrm{Det}\ Y)^{\frac{1}{2}(s_2-s_1)+\frac{1}{4}}\ \zeta(q,l,\chi_1,Y,s)\ ,$$

$$(108)\qquad \lambda(q,l,\chi_1,Y,s) =$$

$$F^*(q,\chi_1,s_2-s_1+\tfrac{1}{2})(\mathrm{Det}\ Y)^{\frac{1}{2}(s_2-s_1)+\frac{1}{4}}\ \zeta^*(q,l,\chi_1,Y,s).$$

From (92), (108) we deduce

$$(109)\qquad \lambda(q,l,\chi_1,Y,s) =$$

$$F^*(q,\chi_1,s_2-s_1+\tfrac{1}{2})(\mathrm{Det}\ Y)^{-\frac{1}{2}(s_1+s_2)}\Lambda(q,l,\chi_1,Y,s)\ .$$

The formulas (80), (81), (95) give us

$$(110)\qquad \lambda(q,l,\chi_1,\tilde{Y},\tilde{s}) = \lambda(q,l,\chi_1,Y,s)\ .$$

From (100) we get

$$(111)\quad \lambda(q,l,\chi_1,Y,s) = (\mathrm{Det}\ Y)^{\frac{1}{2}(s_2-s_1)}\ \frac{1}{2}\int_0^\infty \theta(q,l,\chi_1,Y,t)t^{s_2-s_1}\frac{dt}{t}\ .$$

THEOREM 12: Let $m,p,\epsilon\in\mathbb{R}$; $m,\epsilon>0$ and set

$$(112)\qquad I(m,p) = \int_m^\infty u^p\exp(-u)\frac{du}{u}\ .$$

Then there exists a constant $c_5 = c_5(p,\epsilon) \geq 1$ with

$$(113) \qquad I(m,p) \leq c_5(m^0 + m^{p-\epsilon} + m^{p+\epsilon})$$

PROOF: Obviously

$$(114) \qquad I(m,p) \leq I(1,p) \qquad (m \geq 1)$$

$$(115) \qquad I(m,p) \leq I(1,p) + \Phi(m,p) \qquad (0 < m < 1)$$

with

$$(116) \qquad \Phi(m,p) = \int_m^1 u^{p-1}du \qquad (0 < m < 1)$$

If $p \neq 0$

$$(117) \qquad \Phi(m,p) = [\frac{u^p}{p}]_m^1 \leq \frac{1}{\text{abs } p} + \frac{m^p}{\text{abs } p} \quad .$$

For $\text{abs } p \geq \frac{\epsilon}{2}$ obviously

$$(118) \qquad \Phi(m,p) \leq \frac{2}{\epsilon}(1 + m^p) \qquad (\text{abs } p \geq \frac{\epsilon}{2}) \; .$$

Now

$$(119) \qquad 1 \leq \frac{1}{2}(u^{-\epsilon} + u^{\epsilon}) \; ,$$

and therefore

$$(120) \qquad \Phi(m,p) \leq \frac{1}{2}(\Phi(m,p-\epsilon) + \Phi(m,p+\epsilon)) \; .$$

But for $\text{abs } p \leq \frac{\epsilon}{2}$ we have $\text{abs}(p \pm \epsilon) \geq \frac{\epsilon}{2}$. So from (118), (120) we deduce

$$(121) \qquad \Phi(m,p) \leq \frac{1}{\epsilon}(2 + m^{p-\epsilon} + m^{p+\epsilon}) \qquad (0 < m < 1) \; .$$

This holds for $\text{abs } p \leq \frac{\epsilon}{2}$. But because of (118), (119) it also hold's for abs $p \geq \frac{\varepsilon}{2}$, i. e. for all p. From (114), (115), (121) we deduce (113). Theorem 12 is proved.

THEOREM 13: Let χ_1 be a primitive even character mod q and

$$(122)\quad \lambda_1(q,\mathfrak{l},\chi_1,Y,s) = (\mathrm{Det}\ Y)^{\frac{1}{2}(s_2-s_1)}\ \frac{1}{2}\int_1^\infty \theta(q,\mathfrak{l},\chi_1,Y,t)t^{s_2-s_1}\frac{dt}{t}\ .$$

The integral of the right-hand side of (122) converges absolutely and is a holomorphic function for all $s \in \mathbb{C}^2$. Let R_2 be a non-singular rational 2×2 matrix, $j(Y)$ a positive number with $Y \geq j(Y)E$ and $\mathfrak{K} \subset \mathbb{C}^2$ a compact domain. Then there exists a real number $c_6 = c_6(R_2,\mathfrak{K}) \geq 1$ and three linear functions

$$(123)\qquad \mathcal{L}(\iota,\sigma) = \begin{cases} \pm(\sigma_2-\sigma_1) + j(\iota) \\ \text{or} \\ j(\iota) \end{cases} \qquad (\iota = 1,2,3)$$

with rational $j(\iota)$ $(\iota = 1,2,3)$, such that for $s \in \mathfrak{K}$ the inequality

$$(124)\quad \mathrm{abs}\ \lambda_1(q,\mathfrak{l},\chi_1,Y[R_2],s) \leq$$
$$c_6(\mathrm{Det}\ Y)^{\frac{1}{2}(\sigma_2-\sigma_1)+\frac{1}{4}}\sum_{\iota=1}^{3}(j(Y))^{\mathcal{L}(\iota,\sigma)}$$

holds. By the formula

$$(125)\quad \lambda(q,\mathfrak{l},\chi_1,Y,s) = \lambda_1(q,\mathfrak{l},\chi_1,Y,s) + G(\chi_1)q^{s_1-s_2}\lambda_1(q,\mathfrak{l},\bar{\chi}_1,\check{Y},-s)$$

the function $\lambda(q,\mathfrak{l},\chi_1,Y,s)$ is holomorphically continued to $\mathbb{C}^2$. There is a real number $c_7 = c_7(R_2,\mathfrak{K}) \geq 1$ such that for $s \in \mathfrak{K}$ one has

$$(126)\qquad \mathrm{abs}\ \lambda(q,\mathfrak{l},\chi_1,Y[R_2],s) \leq$$
$$c_6(\mathrm{Det}\ Y)^{\frac{1}{2}(\sigma_2-\sigma_1)+\frac{1}{4}}\sum_{\iota=1}^{3}(j(Y))^{\mathcal{L}(\iota,\sigma)} + c_7(\mathrm{Det}\ Y)^{\frac{1}{2}(\sigma_2-\sigma_1)-\frac{1}{4}}\sum_{\iota=1}^{3}(j(\check{Y}))^{\mathcal{L}(\iota,-\sigma)}.$$

Finally one has the functional equation

(127) $$\lambda(q,1,\chi_1,Y,s) = G(\chi_1)q^{s_1-s_2}\lambda(q,1,\bar{\chi}_1,\check{Y},-s) .$$

PROOF: From theorem 6 and (122) we deduce

$$\text{abs}\ \lambda_1(q,1,\chi_1,Y[R_2],s) \le d_1 j(Y)^{-1}(\text{Det}\ Y)^{\frac{1}{2}(\sigma_2-\sigma_1)+\frac{1}{4}} \times \int_1^\infty t^{\sigma_2-\sigma_1-\frac{1}{2}} \exp(-c_2^{-1}j(Y)t)\frac{dt}{t}$$

with a constant $d_1 \ge 1$. Using $t^* = c_2^{-1}j(Y)t$ as a new variable we get

$$\text{abs}\ \lambda_1(q,1,\chi_1,Y[R_2],s) \le d_1 j(Y)^{\sigma_1-\sigma_2-\frac{1}{2}}(\text{Det}\ Y)^{\frac{1}{2}(\sigma_2-\sigma_1)+\frac{1}{4}} \times I(c_2^{-1}j(Y),\sigma_2-\sigma_1-\tfrac{1}{2}) .$$

From this and theorem 12 we get the first part of theorem 13 till formula (124).

From (111) we conclude

$$\lambda(q,1,\chi_1,Y,s) = \lambda_1(q,1,\chi_1,Y,s)+(\text{Det}\ Y)^{\frac{1}{2}(s_2-s_1)}\frac{1}{2}\int_0^1 \theta(q,1,\chi_1,Y,t)t^{s_2-s_1}\frac{dt}{t}$$

Applying (12), (25) we get

$$\lambda(q,1,\chi_1,Y,s)= \lambda_1(q,1,\chi_1,Y,s)+G(\chi_1)q^{s_1-s_2} \times (\text{Det}\ \check{Y})^{\frac{1}{2}(s_1-s_2)}\int_0^1 \theta(q,1,\bar{\chi}_1,\check{Y},t^{-1})t^{s_2-s_1}\frac{dt}{t} .$$

Making the substitution $t \to t^{-1}$ gives (125).

(126), (127) follow from (18), (124), (125). Theorem 13 is proved.

From (68), (106), (127) we obtain

(128) $$\hat{\lambda}(q,1,\chi_1,Y,\omega) = G(\chi_1)q^{\frac{1}{2}-\omega}\hat{\lambda}(q,1,\bar{\chi}_1,\check{Y},1-\omega).$$

$\hat{\lambda}(q,1,\chi_1,Y,\omega)$ is holomorphic for $\omega \in \mathbb{C}$.

THEOREM 14: Let $q > 1$ and χ_1 an even primitive character mod q. Then $\zeta^*(q,1,\chi_1,Y,s)$ and $\Lambda(q,1,\chi_1,Y,s)$ are holomorphic in the domain

(129) $$\sigma_2 - \sigma_1 \geq 0 .$$

$\zeta(q,1,\chi_1,Y,s)$ is holomorphic for $s \in \mathbb{C}^2$.

PROOF: From $q > 1$ and (104) we deduce $F^*(q,\chi_1,\omega) = \xi(\chi_1,\omega)$. It follows from Landau [22], § 128 that $\xi(\chi,\omega)$ has no zeros for $\operatorname{Re}\omega \geq \frac{1}{2}$. Hence the first part of the theorem follows from (108) (109). The last part is a consequence of (107). Theorem 14 is proved.

THEOREM 15: Let $q > 1$ and χ_1 be an even primitive character mod q. Then $\hat{\zeta}^*(q,1,\chi_1,Y,\omega)$ is holomorphic in the half-plane $\operatorname{Re}\omega \geq \frac{1}{2}$. The function $\hat{\zeta}(q,1,\chi_1,Y,\omega)$ is holomorphic for all $\omega \in \mathbb{C}$.

PROOF: Apply theorem 14.

Set

(130) $$P(1) = q^{\frac{1}{2}} WQ^{-1}(1) = \begin{pmatrix} 0 & q^{-\frac{1}{2}} \\ q^{\frac{1}{2}} & 0 \end{pmatrix} .$$

Then

(131) $$\operatorname{abs} P(1) = 1$$

and

(132) $$P^2(1) = E .$$

From (11), (80) we get

(133) $$\check{\check{Y}} = q^{-1}Y[P(1)] .$$

By theorem 11 and (106) $\lambda(q,1,\chi_1,Y,s)$ is homogeneous in Y of degree 0. Therefore (110), (127) give us

$$(134)\quad \lambda(q,1,\chi_1,Y,(s_1,s_2)) = G(\chi_1)q^{s_1-s_2}\lambda(q,1,\overline{\chi}_1 Y[P(1)],(s_2,s_1)) .$$

This is equivalent to

$$(135)\quad \hat{\lambda}(q,1,\chi_1,Y,\omega) = G(\chi_1)q^{\frac{1}{2}-\omega}\hat{\lambda}(q,1,\overline{\chi}_1,Y[P(1)],1-\omega) .$$

Let $q = 1$. Then $\Psi = \Omega(2)$ and $W, Q(1), P(1) \in \Omega(2)$. Then from (68), theorem 8, (1o7), (108), (110), (127), (128), (134), (135) we deduce

$$(136)\quad \lambda(1,1,1,Y,s) = \lambda(1,1,1,Y,\tilde{s}) = \lambda(1,1,1,Y,(s_2,s_1)) = \lambda(1,1,1,Y,-s) = \lambda(1,1,1,Y^{-1},s),$$

$$(137)\quad \hat{\lambda}(1,1,1,Y,\omega) = \hat{\lambda}(1,1,1,Y,1-\omega) = \hat{\lambda}(1,1,1,Y^{-1},\omega) .$$

§ 3. ELEMENTARY EISTENSTEIN SERIES

§ 3. contains the definition, analytic continuation and functional equation of elementary Eisenstein series.

Let $z \in \mathfrak{Z}(1)$. Like in (41) put

$$(138)\quad Y = \frac{1}{y}\begin{pmatrix} |z|^2 & x \\ x & 1 \end{pmatrix} .$$

Then

$$(139)\quad \mathrm{Det}\, Y = 1 .$$

From (46) we deduce

$$(140)\quad Y[h] = \frac{1}{y}|h_1 z + h_2|^2 .$$

Let $\Gamma(1) = Sp(1,\mathbb{Z})$ and

(141) $$\Gamma_1(1) = \{M = \begin{pmatrix} \pm 1 & b \\ 0 & \pm 1 \end{pmatrix} \in \Gamma(1)\} .$$

For $M = \begin{pmatrix} a & b \\ c & d \end{pmatrix} \in Sp(1,\mathbb{R})$ let, like in (38),

(142) $$M\langle z\rangle = \frac{az+b}{cz+d} = x_M + iy_M$$

with real x_M, y_M. Then

(143) $$y_M = \frac{y}{|cz+d|^2} .$$

Furthermore put

(144) $$M\{z\} = cz + d .$$

With $r \in \mathbb{N} \cup 0$ and $s \in \mathbb{C}$ set

(145) $$\mathcal{E}(1,r,z,s) = \sum_{M \in \Gamma_1(1)\backslash\Gamma(1)} (M\{\bar{z}\})^{2r}\, y_M^{\,s+\frac{1}{2}+r} .$$

Then from (64), (140) we deduce

(146) $$\mathcal{E}(1,0,z,s) = \hat{\zeta}^*(1,1,1,Y,s+\tfrac{1}{2}) .$$

Because of (105), (139), (146) we get

(147) $$\hat{\lambda}(1,1,1,Y,s+\tfrac{1}{2}) = F(s+\tfrac{1}{2})\,\mathcal{E}(1,0,z,s) .$$

From (137), (147) we obtain

(148) $$\hat{\lambda}(1,1,1,Y,s+\tfrac{1}{2}) = \hat{\lambda}(1,1,1,Y,-s+\tfrac{1}{2}) .$$

THEOREM 16: Set $\sigma = \mathrm{Re}\, s$. Let R_2 be a non-singular rational 2×2 matrix, $\mathfrak{k} \subset \mathbb{C}$ compact, $s \in \mathfrak{k}$, $z \in \mathfrak{f}(1)$. Then there exists a real number $c_8 = c_8(R_2,\mathfrak{k}) \geq 1$ and six linear functions

$$(149) \qquad \mathcal{L}(\iota,\sigma) = \begin{cases} \pm\,\sigma + j(\iota) \\ \text{or} \\ j(\iota) \end{cases} \qquad (\iota = 1,2,3,4,5,6).$$

with

$$(150) \qquad \operatorname{abs} \hat{\lambda}(1,1,1,Y[R_2],s+\tfrac{1}{2}) \le c_8 \sum_{\iota=1}^{6} y^{\mathcal{L}(\iota,\sigma)} .$$

PROOF: Because of $z \in \mathcal{F}(1)$ and (43) we may apply theorem 13 with $j(Y) = j(\tilde{Y}) = dy^{-1}$ with some constant d. Now theorem 16 follows from (42) and theorem 13.

It is well known from the theory of elliptic modular forms that the series (145) converges absolutely for $\sigma > \frac{1}{2}$. The case $r = 0$ was just treated.

Now let $r \in \mathbb{N}$. Put

$$(151) \qquad \xi(r,s) = \pi^{-s}\Gamma(s+r)\zeta(2s) .$$

THEOREM 17: Let $\sigma > \frac{1}{2}$ and set

$$(152) \qquad \lambda^*(1,r,z,s) = \xi(r,s+\tfrac{1}{2})\,\mathcal{E}(1,r,z,s) .$$

Then

$$(153) \qquad \lambda^*(1,r,z,s) = \frac{1}{2}\pi^r \int_0^\infty \theta^*(r,z,t)t^{s+\frac{1}{2}+r}\frac{dt}{t} .$$

PROOF: From (48) we obtain

$$\frac{1}{2}\pi^r \int_0^\infty \theta^*(r,z,t)t^{s+\frac{1}{2}+r}\frac{dt}{t} =$$

$$= \frac{1}{2}\pi^r \sum_{h\in\mathbb{Z}^2}(h_1\bar{z}+h_2)^{2r}\int_0^\infty t^{s+\frac{1}{2}+r}\exp(-\tfrac{\pi}{y}|h_1z+h_2|^2t)\frac{dt}{t} =$$

$$\Gamma(s+\tfrac{1}{2}+r)\pi^{-s-\frac{1}{2}}\,\frac{1}{2}\sum_{h\in\mathbb{Z}^2}(h_1\bar{z}+h_2)^{2r}\left(\frac{y}{|h_1z+h_2|^2}\right)^{s+\frac{1}{2}+r} =$$

$$\xi(r,s+\tfrac{1}{2})\tfrac{1}{2}\sum_{\langle c,d\rangle=1}(c\bar z+d)^{2r}\left(\frac{y}{|cz+d|^2}\right)^{s+\frac{1}{2}+r}=\lambda^*(1,r,z,s)\ .$$

Theorem 17 is proved.

From (145) we deduce

$$(154)\qquad \mathcal{E}(1,r,M\langle z\rangle,s) = (M\{\bar z\})^{-2r}\,\mathcal{E}(1,r,z,s) \quad (M\in\Gamma(1)).$$

Hence by (152)

$$(155)\qquad \lambda^*(1,r,M\langle z\rangle,s) = (M\{\bar z\})^{-2r}\lambda^*(1,r,z,s) \quad (M\in\Gamma(1))\ .$$

THEOREM 18: Let $r\in\mathbb{N}$. The function

$$(156)\qquad \lambda_1^*(1,r,z,s) = \frac{1}{2}\,\pi^r\int_1^\infty \theta^*(r,z,t)t^{s+\frac{1}{2}+r}\,\frac{dt}{t}$$

is integral in s and satisfies the equation

$$(157)\qquad \lambda_1^*(1,r,M\langle z\rangle,s) = (M\{\bar z\})^{-2r}\lambda_1(1,r,z,s) \quad (M\in\Gamma(1)).$$

There are a constant $c_9 = c_9(M) > 1$ and three functions $\mathcal{L}(\iota,\sigma)$ of type (149) with

$$(158)\qquad \mathrm{abs}\ \lambda_1^*(1,r,M\langle z\rangle,s) \le c_9\sum_{\iota=1}^{3} y^{\mathcal{L}(\iota,\sigma)} \quad (z\in\mathfrak{F}(1),\ M\in\Gamma(1)).$$

By

$$(159)\qquad \lambda^*(1,r,z,s) = \lambda_1^*(1,r,z,s) + \lambda_1^*(1,r,z,-s)$$

the function $\lambda^*(1,r,z,s)$ is holomorphically continued to $\mathbb{C}$ and it satisfies the functional equation

$$(160)\qquad \lambda^*(1,r,z,s) = \lambda^*(1,r,z,-s)\ .$$

Furthermore there is an estimation

$$(161)\qquad \mathrm{abs}\ \lambda^*(1,r,M\langle z\rangle,s) \le c_9\sum_{\iota=1}^{6} y^{\mathcal{L}(\iota,\sigma)} \quad (z\in\mathfrak{F}(1),\ M\in\Gamma(1)).$$

with functions $\mathcal{L}(\iota,\sigma)$ of type (149).

PROOF: (157) follows from (54). Let $z \in \mathcal{f}(1)$, $M \in \Gamma(1)$. From theorems 7 and 12 we get

$$\operatorname{abs} \lambda_1^*(1,r,M\langle z\rangle,s) \le c_9 y^{\sigma+\frac{1}{2}+r} \int_1^\infty \left(\frac{t}{y}\right)^{\sigma-\frac{1}{2}} \exp\left(-c_4^{-1}\frac{t}{y}\right)\frac{dt}{t}$$

$$= c_9 y^{\sigma+\frac{1}{2}+r} I\left(y^{-1},\sigma-\tfrac{1}{2}\right) .$$

Now the estimate (158) follows from theorem 12.

Because of (158) the function $\lambda_1^*(1,r,z,s)$ is integral in s .

From (53), (153) we get

$$\lambda^*(1,r,z,s) = \lambda_1^*(1,r,z,s) + \tfrac{1}{2}\pi^r \int_0^1 \theta^*(r,z,t)t^{s+\frac{1}{2}+r}\frac{dt}{t}$$

$$= \lambda_1^*(1,r,z,s) + \tfrac{1}{2}\pi^r \int_0^1 \theta^*(r,z,t^{-1})t^{s-\frac{1}{2}-r}\frac{dt}{t} =$$

$$\lambda_1^*(1,r,z,s) + \tfrac{1}{2}\pi^r \int_1^\infty \theta^*(r,z,t)t^{-s+\frac{1}{2}+r}\frac{dt}{t} =$$

$$\lambda_1^*(1,r,z,s) + \lambda_1^*(1,r,z,-s) .$$

Herewith we have (159). From (158), (159) we obtain (160), (161). Theorem 18 is proved.

CHAPTER II. PREPARATIONAL MATERIAL

In the following two chapters we develop the theory of Selberg's zetafunctions and L-series. In the present chapter we collect some material which will be needed later. We prove some results on systems of Dirichletcharacters. We investigate matrices and subgroups of the unimodular group. We collect basic results of the space of positive matrices and on thetafunctions.

----

§ 4. SYSTEMS OF PRIMITIVE CHARACTERS

In this paragraph we investigate the following problem. Let $\chi_1,\ldots,\chi_m$ denote m characters mod q. Under with assumptions are all products $\chi_\mu \cdots \chi_\nu$ $(1 \leq \mu \leq \nu \leq m)$ primitive?

In this paragraph we shall investigate the following problem. Let $m,q \in \mathbb{N}$, $q > 1$. Under which supposition exist m characters $\chi_1,\ldots,\chi_m$ mod q such that the products

(162) $$\overset{o}{\chi}_{\nu\mu} = \chi_\mu\cdots\chi_\nu \qquad (1 \leq \mu \leq \nu \leq m)$$

are primitive? Especially we are interested in the case that $\chi_1,\ldots,\chi_m$ are even.

THEOREM 19: Consider the decomposition

(163) $$q = q_1\cdots q_r$$

with

(164) $$\langle q_\iota, q_\varkappa\rangle = 1 \qquad (\iota \neq \varkappa;\ \iota,\varkappa = 1,\ldots,r).$$

For $a \in \mathbb{Z}$, $\langle a,q\rangle = 1$ define the residueclasses a_ι mod q $(\iota = 1,\ldots,r)$ by

(165) $$a_\iota \equiv a \bmod q_\iota,\ a_\iota \equiv 1 \bmod \frac{q}{q_\iota} \quad (\iota = 1,\ldots,r)\ .$$

Let χ be a character mod q and define the characters $\chi^{(\iota)}$ $(\iota = 1,\ldots,r)$ by

(166) $$\chi^{(\iota)}(a) = \chi(a_\iota) \qquad (\iota = 1,\ldots,r)\ .$$

Then $\chi^{(\iota)}$ is a character mod q_ι $(\iota = 1,\ldots,r)$ and one has

(167) $$a \equiv a_1 \cdots a_r \bmod q\ ,$$

(168) $\chi(a) = \chi(a_1) \ldots \chi(a_r) = \chi^{(1)}(a) \ldots \chi^{(r)}(a)$.

The formulas (166) and (168) define a bijective map between χ and the system

(169) $$\chi^{(1)}, \ldots, \chi^{(r)} .$$

χ is a primitive character mod q if and only if $\chi^{(\iota)}$ is a primitive character mod q_ι $(\iota = 1,\ldots,r)$. Furthermore χ is even if and only if there is an even number of odd characters among $\chi^{(1)},\ldots,\chi^{(r)}$. Especially χ is even if all $\chi^{(1)},\ldots,\chi^{(r)}$ are even.

Let $\chi_1,\ldots,\chi_m$ be characters mod q and $\mathring{\chi}_{\nu\mu}$ defined by (162). According (166) define the characters $\chi_\mu^{(\iota)}$, $\mathring{\chi}_{\nu\mu}^{(\iota)}$ $(1 \leq \mu \leq \nu \leq m;$ $\iota = 1,\ldots,r)$. Then

(170) $$\mathring{\chi}_{\nu\mu}^{(\iota)} = \chi_\mu^{(\iota)} \ldots \chi_\nu^{(\iota)} \quad (1 \leq \mu \leq \nu \leq m;\ \iota = 1,\ldots,r).$$

PROOF: The assertion about the even characters follows from (168) the rest from Hasse [15], § 13,6, especially pages 210, 211.

By theorem 19 our problem is reduced to the case that q is a power of a prime-number.

THEOREM 20: Let p be a prime-number, $k \in \mathbb{N}$ and

(171) $$q = p^k .$$

For $p = 2$, $k = 1$ there exists no primitive character. For $p = 2$, $k \geq 2$ each character χ mod 2^k may be uniquely written as

(172) $$\chi = \chi_4^c\, \chi_{2^k}^d \quad (c \bmod 2,\ d \bmod 2^{k-2}) .$$

Here χ_4, χ_{2^k} are "basischaracters" with the conductors 4 respectively 2^k For $k = 2$ the factor $\chi_{2^k}^d$ does not appear. χ is a primitive character mod 2^k, if and only if

(173) $$c \not\equiv 0 \bmod 2 \qquad (k = 2),$$

(174) $$d \not\equiv 0 \bmod 2 \qquad (k > 2).$$

It is

(175) $$\chi_4(-1) = -1$$

(176) $$\chi_{2^k}(-1) = 1 \ .$$

Hence χ is even if and only if

(177) $$c \equiv 0 \bmod 2 \ .$$

For $p \geq 3$, $k \geq 1$ each character χ mod p^k may be uniquely written as

(178) $$\chi = \chi_p^c \, \chi_{p^k}^d \quad (c \bmod p-1,\ d \bmod p^{k-1}) \ .$$

Here χ_p and χ_{p^k} are "basischaracters" with the conductors p respectively p^k.

For $k = 1$ the factor $\chi_{p^k}^d$ does not appear. χ is a primitive character mod p^k if and only if

(179) $$c \not\equiv 0 \bmod p-1 \qquad (k = 1) \ ,$$

(180) $$d \not\equiv 0 \bmod p \qquad (k > 1) \ .$$

It is

(181) $$\chi_p(-1) = -1$$

(182) $$\chi_{p^k}(-1) = 1 \ .$$

χ is even if and only if (177) is true.

PROOF: All assertions except the formulas (181), (182) may be found in Hasse [15], § 13.6, page 212. For the proof of the formulas (181), (182) we use the representation

$$a \equiv w^{\alpha'}(1+p)^{\alpha''} \bmod p^k \tag{183}$$

with

$$0 \leq \alpha' < p-1,\ 0 \leq \alpha'' < p^{k-1} \quad (p \geq 3) \tag{184}$$

from Hasse [15], page 212 third row from above. We set $a = -1$ and take the square of (183). Then

$$1 \equiv w^{2\alpha'}(1+p)^{2\alpha''} \bmod p^k , \tag{185}$$

i. e.,

$$2\alpha' \equiv 0 \bmod p-1,\ 2\alpha'' \equiv 0 \bmod p^{k-1} . \tag{186}$$

From (184), (186) we get

$$\alpha' = \frac{p-1}{2} ,\ \alpha'' = 0 . \tag{187}$$

Hence

$$-1 \equiv w^{\frac{p-1}{2}}(1+p)^0 \bmod p^k. \tag{188}$$

Because of (188) the formulas (181), (182) follow from the left box in Hasse [15], middle of page 212. Theorem 20 is proved.

Choose characters $\psi_1,\dots,\psi_{m+1}$ with

$$\psi_{\nu+1}^{-1}\ \psi_\nu = \chi_\nu \qquad (\nu = 1,\dots,m) . \tag{189}$$

Then

$$\psi_{\nu+1}^{-1}\ \psi_\mu = \mathring{\chi}_{\nu\mu} \qquad (1 \leq \mu \leq \nu \leq m). \tag{190}$$

THEOREM 21: For

(191) $$q = 2^k \qquad (k \geq 1)$$

the $\mathring{\chi}_{\nu\mu}$ $(1 \leq \mu \leq \nu \leq m)$ can be primitive only for $m=1$. For $k = 1$ there is no primitive character, for $k = 2$ there is only one odd primitive character, for $k > 2$ there are even and odd p primitive characters.

PROOF: For $k = 1$ the assertion follows from theorem 20. For $k \geq 2$ apply (190) and set

(192) $$\psi_\mu = \chi_4^{c_\mu} \chi_{2^k}^{d_\mu} \qquad (\mu = 1,\dots,m+1) .$$

By theorem 20 the $\mathring{\chi}_{\nu\mu}$ $(1 \leq \mu \leq \nu \leq m)$ are primitive if and only if

(193) $$c_\mu \not\equiv c_\nu \bmod 2 \qquad (1 \leq \mu < \nu \leq m+1)\ (k = 2),$$

(194) $$d_\mu \not\equiv d_\nu \bmod 2 \qquad (1 \leq \mu < \nu \leq m+1)\ (k > 2)$$

holds. χ_μ $(\mu = 1,\dots,m)$ is even, if

(195) $$c_{\mu+1} \equiv c_\mu \bmod 2 \qquad (\mu = 1,\dots,m)$$

From this theorem 21 follows immediately.

THEOREM 22: Let p be an odd prime, $k \in \mathbb{N}$ and

(196) $$q = p^k .$$

If there are characters $\chi_1,\dots,\chi_m$ such that all $\mathring{\chi}_{\nu\mu}$ $(1 \leq \mu \leq \nu \leq m)$ are primitive, then

(197) $$m < p-1 \qquad (k = 1),$$

(198) $$m < p \qquad (k > 1) .$$

Now let (197) respectively (198) be fulfilled and let $\varkappa$ be an arbitrary index with $1 \leq \varkappa \leq m$. Furthermore let χ be a primitive character mod q. Then there are characters $\chi_1, \ldots, \chi_m$, such that all $\overset{\circ}{\chi}_{\nu\mu}$ $(1 \leq \mu \leq \nu \leq m)$ are primitive and that furthermore

(199) $$\chi_\varkappa = \chi$$

holds. In case $k > 1$ moreover each character χ_μ $(\mu \neq \varkappa;\ \mu = 1, \ldots, m)$ may be chosen even or odd. If one demands that all characters $\chi_1, \ldots, \chi_m$ are even the same assertion is true if the condition (197) is replaced by

(200) $$m < \frac{p-1}{2} \quad .$$

PROOF: Set

(201) $$\psi_\mu = \chi_p^{c_\mu} \chi_{p^k}^{d_\mu} \qquad (\mu = 1, \ldots, m+1).$$

By theorem 20 the $\overset{\circ}{\chi}_{\nu\mu}$ $(1 \leq \mu \leq \nu \leq m)$ are primitive, if and only if

(202) $$c_\mu \not\equiv c_\nu \bmod p-1 \qquad (1 \leq \mu < \nu \leq m+1)\ (k = 1)$$

(203) $$d_\mu \not\equiv d_\nu \bmod p \qquad (1 \leq \mu < \nu \leq m+1)\ (k > 1)$$

holds. The χ_μ $(\mu = 1, \ldots, m)$ are even if

(204) $$c_{\mu+1} \equiv c_\mu \bmod 2 \quad (\mu = 1, \ldots, m) \ .$$

PROOF: Consider first the case $k = 1$. Let (202) be true. Since there are exactly $p-1$ different residueclasses mod $p-1$ the inequality (197) follows. Now let (197) be true and set

(205) $$\chi = \chi_p^a \chi_{p^k}^b \quad .$$

Then (202) may be fulfilled with the additional condition

(206) $$c_{\varkappa+1} - c_\varkappa \equiv a \bmod p-1 \ .$$

Hence the $\overset{\circ}{\chi}_{\nu\mu}$ $(1 \leq \mu \leq \nu \leq m)$ are primitive and (199) holds. To demand that all characters $\chi_1,\ldots,\chi_m$ are even is the same as to demand that all $\psi_1,\ldots,\psi_{m+1}$ are even, i. e.,

(207) $$c_\mu \equiv 0 \bmod 2 \qquad (\mu = 1,\ldots,m+1)\ .$$

Then the same as before is true with (200) instead of (197).

Now consider the case $k > 1$. Then (203) may be fulfilled if and only if (198) holds. If (198) is true one may in addition to (203) obtain (206) and

(208) $$d_{\varkappa+1} - d_\varkappa \equiv b \bmod p\ .$$

Hence (199) holds. Except of the condition (206) the c_μ may be chosen arbitrary. Hence χ_μ $(\mu \neq \varkappa)$ may be chosen even or odd. Theorem 22 is proved.

THEOREM 23: Let

(209) $$q = 2^t p_1 \ldots p_s p_{s+1}^{t_{s+1}} \ldots p_r^{t_r}\ .$$

Here $p_1,\ldots,p_r$ are different odd primes,

(210) $$p_1 < p_2 < \ldots < p_s\ ;$$

(211) $$p_{s+1} < p_{s+2} < \ldots < p_r\ ,$$

(212) $$0 \leq t\ ,$$

(213) $$t_{s+1},\ldots,t_r \geq 2\ .$$

If there are characters $\chi_1,\ldots,\chi_m$ mod q, such that all the products $\overset{\circ}{\chi}_{\nu\mu} = \chi_\mu \cdots \chi_\nu$ $(1 \leq \mu \leq \nu \leq m)$ are primitive, then either $t \geq 2$, $m = 1$ or

(214) $$t = 0,\ m < \min(p_1-1, p_{s+1})\ .$$

Now let (214) be true, let χ be an arbitrary primitive character mod q and $\varkappa$ an arbitrary index with $1 \leq \varkappa \leq m$. Then there are characters $\chi_1,\ldots,\chi_m$ mod q, such that all products $\mathring{\chi}_{\nu\mu} = \chi_\mu \cdots \chi_\nu$ $(1 \leq \mu \leq \nu \leq m)$ are primitive and moreover $\chi_\varkappa = \chi$. Let

$$t = 0, \ m < \min\left(\frac{p_1 - 1}{2}, \ p_s + 1\right), \tag{215}$$

$\varkappa$ an arbitrary index with $1 \leq \varkappa \leq m$ and χ an even primitive character mod q. Then there are even characters $\chi_1,\ldots,\chi_m$ mod q, such that all products $\mathring{\chi}_{\nu\mu}$ $(1 \leq \mu \leq \nu \leq m)$ are primitive and furthermore $\chi_\varkappa = \chi$.

PROOF: Combine theorems 19, 21, 22 .

§ 5. MATRICES

In this paragraph we prove results on matrices with integral coefficients and we consider subgroups of the unimodular group

---

Let $n,w,q \in \mathbb{N}$,

$$n > 1, \ w > 1, \ q \geq 1, \tag{216}$$

$$k_o = 0, \ k_\iota \in \mathbb{N} \ (\iota = 1,\ldots,w), \tag{217}$$

$$0 = k_o < k_1 < \ldots < k_w = n, \tag{218}$$

$$l_\nu = k_\nu - k_{\nu-1} \quad (\nu = 1,\ldots,w), \tag{219}$$

$$k = (k_1,\ldots,k_w), \ l = (l_1,\ldots,l_w). \tag{220}$$

Then

$$k_\nu = \sum_{\iota=1}^{\nu} l_\iota \qquad (\nu = 1,\ldots,w). \tag{221}$$

Especially let

$$\overset{\circ}{k} = (1,2,\dots,n)\ ,\ \overset{\circ}{l} = (1,\dots,1) \tag{222}$$

with n times 1. Set

$$\tilde{l}_\nu = l_{w+1-\nu} \qquad (\nu = 1,\dots,w)\ , \tag{223}$$

$$\tilde{l} = (\tilde{l}_1,\dots,\tilde{l}_w)\ , \tag{224}$$

$$\tilde{k}_o = 0,\ \tilde{k}_\nu = \sum_{\iota=1}^{\nu} \tilde{l}_\iota \ (\nu = 1,\dots,w), \tag{225}$$

$$\tilde{k} = (\tilde{k}_1,\dots,\tilde{k}_w)\ . \tag{226}$$

Then

$$\tilde{k}_\nu = k_w - k_{w-\nu} = n - k_{w-\nu} \quad (\nu = 0,\ .\ ,w) \tag{227}$$

DEFINITION 1: Let $\iota = 1,\dots,n$ and $\nu = 1,\dots,w$. Then set

$$\omega(\iota) = \nu \tag{228}$$

for

$$k_{\nu-1} < \iota \leq k_\nu\ . \tag{229}$$

Furthermore set

$$\tilde{\omega}(\iota) = \nu \tag{230}$$

for

$$\tilde{k}_{\nu-1} < \iota \leq \tilde{k}_\nu\ . \tag{231}$$

Obviously

$$\tilde{\omega}(\iota) = w+1 - \omega(n+1-\iota) \quad (\iota = 1,\dots,n)\ . \tag{232}$$

Set

(233) $$w^* = w-1, \; n^* = k_{w-1} \, ,$$

(234) $$k^* = (k_1,\ldots,k_{w*}) \; , \; l^* = (l_1,\ldots,l_{w*}) \; .$$

Then

(235) $$n = n^* + l_w \; .$$

Let

(236) $$\check{l}_\nu = l_{w-\nu} \quad (\nu = 1,\ldots,w-1), \; \check{l}_w = l_w \; ,$$

(237) $$\check{l} = (\check{l}_1,\ldots,\check{l}_w) \; ,$$

(238) $$\check{k}_o = 0 \; , \; \check{k}_\nu = \sum_{\iota=1}^{\nu} \check{l}_\iota \quad (\nu = 1,\ldots,w),$$

(239) $$\check{k} = (\check{k}_1,\ldots,\check{k}_w) \; ,$$

(240) $$\check{l} = (\widetilde{l^*},l_w), \quad \check{k} = (\widetilde{k^*},k_w) \; .$$

Let $\mathfrak{z}$ denote the cyclic permutation

(241) $$\mathfrak{z}l = (l_w,l_1,\ldots,l_{w-1}),$$

and set

(242) $$\mathfrak{z}^*l^* = (l_{w*},l_1,\ldots,l_{w*-1}) \; ,$$

(243) $$\hat{l}_1 = l_{w*}, \; \hat{l}_\nu = l_{\nu-1} \; (\nu = 2,\ldots,w^*), \; \hat{l}_w = l_w.$$

Then

(244) $$\hat{l} = (\hat{l}_1,\ldots,\hat{l}_w) = (\mathfrak{z}^*l^*,l_w) \; ,$$

(245) $$\hat{k}_o = 0, \; \hat{k}_\nu = \sum_{\iota=1}^{\nu} \hat{l}_\iota \quad (\iota = 1,\ldots,w) \; .$$

DEFINITION 2: If $n \geq m$ and G is an integral $n \times m$ matrix, $\langle G \rangle \geq 0$ is the greatest common devisor of all $m \times m$ subdeterminants of G. If $\langle G \rangle = 1$ we say that G is "primitive" .

If A is a $k_\beta \times k_\gamma$ matrix $(1 \leq \beta, \gamma \leq w)$ let $A = (\overset{1}{A}_{\nu\mu})$ be the splitting of A in $l_\nu \times l_\mu$ submatrices $\overset{1}{A}_{\nu\mu}$ belonging to l. Set $\overset{1}{A}_\nu = \overset{1}{A}_{\nu\nu}$ $(\nu = 1,\ldots,\min(\beta,\gamma))$. Obviously $\overset{1}{A}_{\nu\mu}$ consits of those elements $a_{\iota\varkappa}$ of A with $\omega(\iota) = \nu$, $\omega(\varkappa) = \mu$.

Let $\Omega(n)$ be the group of unimodular nxn matrices U, i. e. of the integral matrices U with abs $U = 1$. Define the following subgroups of $\Omega(n)$

$$(246) \quad \Psi_1(l) = \{U = (\overset{1}{U}_{\nu\mu}) \in \Omega(n);\ \overset{1}{U}_{\nu\mu} \equiv 0 \bmod q \quad (1 \leq \mu < \nu \leq w)\},$$

$$(247) \quad \Delta_1(l) = \{U = (\overset{1}{U}_{\nu\mu}) \in \Omega(n);\ \overset{1}{U}_{\nu\mu} = 0 \qquad (1 \leq \mu < \nu \leq w)\} ,$$

$$(248) \quad \Psi_2(l) = \Psi_1(\overset{\circ}{l}),\ \Delta_2(l) = \Delta_1(l) \cap \Psi_1(\overset{\circ}{l}) ,$$

$$(249) \quad \Omega(n,q) = \{U \in \Omega(n);\ U \equiv E \bmod q\} .$$

$\Omega(n,q)$ is called the "principal congruence subgroup" of $\Omega(n)$ of level q .

THEOREM 24: It is

$$(250) \quad (\mathrm{Det}\ \overset{1}{U}_1)\ldots(\mathrm{Det}\ \overset{1}{U}_w) \equiv \mathrm{Det}\ U = \pm 1 \bmod q \quad (U \in \Psi_1(l)),$$

$$(251) \quad u_1\ldots u_n \equiv \mathrm{Det}\ U = \pm 1 \bmod q \qquad (U \in \Psi_2(l)),$$

$$(252) \quad \mathrm{Det}\ \overset{1}{U}_\nu = \pm 1 \qquad (\nu = 1,\ldots,w)\ (U \in \Delta_1(l)).$$

PROOF: Clear.

For $1 \leq \varkappa \leq \iota \leq w$ let $\mathfrak{A}_1(1,\iota,\varkappa)$ be the set of integral $k_\iota \times k_\varkappa$ matrices $A = (\overset{1}{A}_{\nu\mu})$ with $\overset{1}{A}_{\nu\mu} \equiv 0 \bmod q$ $(1 \leq \mu \leq \varkappa;\ 1 \leq \mu < \nu \leq \iota)$. Set $\mathfrak{A}_2(1;\iota,\varkappa) = \mathfrak{A}_1(\mathring{1};k_\iota,k_\varkappa)$,

$$(253)\quad \mathfrak{B}_\alpha(1;\iota,\varkappa) = \{A \in \mathfrak{A}_\alpha(1;\iota,\varkappa); \langle\langle A\rangle,q\rangle = 1, \mathrm{Rk}\, A = k_\varkappa\} \quad (\alpha = 1,2),$$

$$(254)\quad \mathfrak{L}_\alpha(1;\iota,\varkappa) = \{A \in \mathfrak{A}_\alpha(1;\iota,\varkappa);\ \langle A\rangle = 1\} \quad (\alpha = 1,2).$$

For $q > 1$, the condition $\mathrm{Rk}\, A = k_\varkappa$ in (253) follows from $\langle\langle A\rangle,q\rangle = 1$.

THEOREM 25: It is

$$(255)\qquad \mathrm{Rk}\, A = k_\varkappa \qquad (A \in \mathfrak{B}_\alpha(1;\iota,\varkappa)) .$$

PROOF: Clear.

Let $\mathfrak{H}_1(1;\iota,\varkappa)$ be the set of integral $k_\iota \times k_\varkappa$ matrices $A = (\overset{1}{A}_{\nu\mu})$ with $\overset{1}{A}_{\nu\mu} = 0$ $(1 \leq \mu \leq \varkappa;\ 1 \leq \mu < \nu \leq \iota)$. Set $\mathfrak{H}_2(1;\iota,\varkappa) = \mathfrak{H}_1(\mathring{1},k_\iota,k_\varkappa)$,

$$(256)\qquad \mathfrak{J}_\alpha(1;\iota,\varkappa) = \mathfrak{H}_\alpha(1;\iota,\varkappa) \cap \mathfrak{B}_\alpha(1;\iota,\varkappa) \qquad (\alpha = 1,2) ,$$

$$(257)\qquad \mathfrak{K}_\alpha(1;\iota,\varkappa) = \mathfrak{H}_\alpha(1;\iota,\varkappa) \cap \mathfrak{L}_\alpha(1;\iota,\varkappa) \qquad (\alpha = 1,2) .$$

Set $A = (a_{\nu\mu})$ $(\nu = 1,\ldots,k_\iota;\ \mu = 1,\ldots,k_\varkappa)$. Then

$$\mathfrak{H}^{\geq}(1;\iota,\varkappa), \mathfrak{J}^{\geq}(1;\iota,\varkappa), \mathfrak{K}^{\geq}(1;\iota,\varkappa),$$

$$\mathfrak{H}^{>}(1;\iota,\varkappa), \mathfrak{J}^{>}(1;\iota,\varkappa), \mathfrak{K}^{>}(1;\iota,\varkappa)$$

denote those subsets of $\mathfrak{H}_2(1;\iota,\varkappa)$, $\mathfrak{J}_2(1;\iota,\varkappa)$, $\mathfrak{K}_2(1;\iota,\varkappa)$, for which $a_\nu \geq 0$ respectively $a_\nu > 0$ $(\nu = 1,\dots,k_\varkappa)$. Finally let $\mathfrak{h}^*(1;\iota,\varkappa)$, $\mathfrak{J}^*(1;\iota,\varkappa)$, $\mathfrak{k}^*(1;\iota,\varkappa)$ be those subsets of $\mathfrak{h}^>(1;\iota,\varkappa)$, $\mathfrak{J}^>(1;\iota,\varkappa)$, $\mathfrak{k}^>(1;\iota,\varkappa)$, for which

(258) $$0 \leq a_{\mu\nu} < a_\mu \qquad (1 \leq \mu < \nu \leq k_\varkappa) .$$

THEOREM 26: In $\mathfrak{h}^*(1;\iota,\varkappa)$, $\mathfrak{J}^*(1;\iota,\varkappa)$, $\mathfrak{k}^*(1;\iota,\varkappa)$ there are exactly

(259) $$a_1^{k_\varkappa-1}\; a_2^{k_\varkappa-2} \cdots a_{k_\varkappa-1}^{1}$$

matrices A with prescribed diagonalelements $a_1,\dots,a_{k_\varkappa}$.

PROOF: Clear.

THEOREM 27: Let P be an integral $n\times m$ matrix with $1 \leq m \leq n$. There exists a matrix $U \in \Omega(n,q)$ with

(260) $$U = (P\ *)$$

if and only if P is primitive and

(261) $$P \equiv \binom{E}{0} \bmod q .$$

Here E denotes the $m\times m$ unit matrix.

PROOF: See Christian [7], page 24, Satz 3.6.

THEOREM 28: Let P be a $n \times k_\varkappa$ matrix $(1 \leq \varkappa \leq w)$. There exists an $U \in \Psi_\alpha(1)$ with

(262) $$U = (P\ *)$$

if and only if $P \in \mathfrak{L}_\alpha(1;w,\varkappa)$ $(\alpha = 1,2)$.

PROOF: From $U = (P\ *) \in \Psi_\alpha(1)$ follows immediately $P \in \mathcal{L}_\alpha(1;w,\varkappa)$. Now let $P \in \mathcal{L}_\alpha(1;w,\varkappa)$. We have to show that there exists a $U \in \Psi_\alpha(1)$ with (262). If $P \in \mathcal{L}_1(1;w,\varkappa)$ there is a $V \in \Delta_1(1)$ with $VP \in \mathcal{L}_2(1;w,\varkappa)$. Therefore it suffices to prove the theorem for $\alpha = 2$, i. e. for $\Psi_1(\mathring{1})$ respectively $\mathcal{L}_1(\mathring{1};n,m)$.

Now let be

$$P \in \mathcal{L}_1(\mathring{1};n,m) \; . \tag{263}$$

For $n = m$ the assertion is true. Now let $1 \leq m < n$. At first we consider the case $m = 1$. Then we have

$$P \equiv (p_1 0 \ldots 0)' \bmod q \; . \tag{264}$$

Chose an $a \in \mathbb{Z}$ with $ap_1 \equiv 1 \bmod q$ and set

$$V_1 = \begin{pmatrix} a & 1 & 0 \\ ap_1 - 1 & p_1 & 0 \\ 0 & 0 & E^{(n-2)} \end{pmatrix} \in \Psi_1(\mathring{1}) \; . \tag{265}$$

Then

$$V_1 P \equiv (1\ 0 \ldots 0)' \bmod q \; . \tag{266}$$

Because of theorem 27 there is a $V_2 \in \Omega(n,q) \subset \Psi_1(\mathring{1})$ with $V_2 V_1 P = (1\ 0\ \ldots\ 0)'$. Set $V = V_2 V_1 \in \Psi_1(\mathring{1})$. Then

$$VP = (1\ 0 \ldots 0)' \; . \tag{267}$$

Now let $m > 1$. We make the induction assumption that for each $P_1 \in \mathcal{L}_1(\mathring{1};n,m-1)$ there exists a $V_1 \in \Psi_1(\mathring{1})$ such that

$$V_1 P_1 = \begin{pmatrix} E^{(m-1)} \\ 0 \end{pmatrix} . \tag{268}$$

From (263) it follows $P = (P_1\ *)$ with $P_1 \in \mathcal{L}_1(\mathring{1},n,m-1)$. Hence there is a $V_1 \in \Psi_1(\mathring{1})$ with

$$V_1 P = \begin{pmatrix} E^{(m-1)} & d \\ 0 & P_2 \end{pmatrix} , \tag{269}$$

(270) $$P_2 \in \mathcal{L}_1(\overset{1}{1}, n+1-m, 1) \ .$$

Here $\overset{1}{1} = (1,\ldots,1)$ with $n+1-m$ times 1. As already proved there is a $V_2^* \in \Psi_1(\overset{1}{1})$ with

(271) $$V_2^* P_2 = (1\ 0\ldots 0)' \ .$$

With

(272) $$V = \begin{pmatrix} E^{(m-1)} & -d & 0 \\ 0 & 1 & 0 \\ 0 & 0 & E^{(n-m)} \end{pmatrix} \begin{pmatrix} E^{(m-1)} & 0 \\ 0 & V_2^* \end{pmatrix} V_1 \in \Psi_1(\overset{\circ}{1})$$

follows

(273) $$VP = \begin{pmatrix} E^{(m)} \\ 0 \end{pmatrix} \ .$$

For each $P \in \mathcal{L}_1(\overset{\circ}{1};n,m)$ there is a $V \in \Psi_1(\overset{\circ}{1})$ with (273). Now (262) follows with $U = V^{-1}$.

Theorem 28 is proved.

Let $\breve{\mathcal{U}}(1)$ be an arbitrary but fixed system of representatives of the cosets $\Psi_2(1)/\Delta_2(1)$ and assume $E \in \breve{\mathcal{U}}(1)$. Then $\breve{\mathcal{U}}(1)$ can be also considered as a complete set of representatives of the cosets $\Psi_1(1)/\Delta_1(1)$.

THEOREM 29: A complete set of representatives of the cosets $\mathcal{B}_\alpha(1;w,w^*)/\Delta_\alpha(1^*)$ $(\alpha = 1,2)$ is given by the products

(274) $$B = UD$$

with

(275) $$U = (\overset{1}{U}_{\nu\mu}) \in \breve{\mathcal{U}}(1) \ ,$$

(276) $$D = (\overset{1}{D}_{\nu\mu}) \in \mathcal{J}^*(1;w,w^*) .$$

Furthermore

(277) $$\overset{1}{B}_\nu \equiv \overset{1}{U}_\nu \overset{1}{D}_\nu \bmod q \quad (\nu = 1,\ldots,w^*) .$$

PROOF: If $B \in \mathcal{B}_1(1;w,w^*)$ there exists a $V \in \Delta_1(1^*)$ with $BV \in \mathcal{B}_2(1;w,w^*)$. Hence it suffices to prove the theorem for $\alpha = 2$.

Now

(278) $$\mathcal{B}_2(1;w,w^*) = \mathcal{B}_1(\mathring{1};m,n^*) .$$

First we prove that each

(279) $$B \in \mathcal{B}_1(\mathring{1};n,m)$$

may be written as

(280) $$B = U_1\ D_1$$

with

(281) $$U_1 \in \Psi_1(\mathring{1}) ,$$

(282) $$D_1 \in \mathcal{J}_1(\mathring{1};n,m) ,$$

At first let $m = 1$ and $a = \langle B\rangle$.

From (253), (279) it follows $\langle a,q\rangle = 1$, hence $B = aP$ with $P \in \mathcal{L}_1(\mathring{1};n,1)$. By theorem 28 there is a $V \in \Psi_1(\mathring{1})$ with $VP = (1\ 0\ldots0)'$. Hence $VB = (a\ 0\ldots0)'$. Therefore for each $B \in \mathcal{B}_1(\mathring{1};n,1)$ there exists a $V \in \Psi_1(\mathring{1})$ with $VB \in \mathcal{J}_1(\mathring{1};n,1)$. Now let $m > 1$. We make the induction assumption that for each $B \in \mathcal{L}_1(\mathring{1};n,m-1)$ there exists a $V \in \Psi_1(\mathring{1})$ with $VB \in \mathcal{J}_1(\mathring{1};n,m-1)$. From (279) follows $B = (B_1\ *)$ with $B_1 \in \mathcal{L}_1(\mathring{1};n,m-1)$. Hence there exists a $V_1 \in \Psi_1(\mathring{1})$ with

$$(283) \qquad V_1 B = \begin{pmatrix} a_1 & * & * \\ & \ddots \ a_{m-1} & \\ 0 & & b_1 \end{pmatrix} .$$

Here

$$(284) \qquad b_1 \in \mathcal{B}_1(\overset{1}{l}, n+1-m, 1) .$$

Hence there is a $V_2 \in \Psi_1(\overset{1}{l})$ with $V_2 b_1 \in \mathcal{J}_1(\overset{1}{l}, n+1-m, 1)$. With

$$(285) \qquad V = \begin{pmatrix} E^{(m-1)} & 0 \\ 0 & V_2 \end{pmatrix} V_1 \in \Psi_1(\overset{\circ}{l})$$

follows

$$(286) \qquad D_1 = VB \in \mathcal{J}_1(\overset{\circ}{l}; n, m) .$$

Setting $U_1 = V^{-1}$ we obtain (280), (281), (282).

Set

$$(287) \qquad W = W(n) = \begin{pmatrix} 0 & & 1 \\ & \cdot^{\cdot^{\cdot}} & \\ 1 & & 0 \end{pmatrix} .$$

Obviously each matrix

$$(288) \qquad B_2 \in \mathcal{B}_2(\overset{\circ}{l}, n, n)$$

may be written as

$$(289) \qquad B_2 = D_2 U_2$$

with

(290) $$U_2 \in \Psi_1(\mathring{1}),$$

(291) $$D_2 \in \mathcal{J}_1(\mathring{1};n,n) ,$$

because if we set

$$B_1 = WB_2'W, \; U_1 = WU_2'W, \; D_1 = WD_2'W ,$$

the conditions (289), (290), (291) are equivalent to (280), (281), (282).

Now proceeding with (280) we set

(292) $$U_1 = UV^*$$

with (275) and

(293) $$V^* \in \Delta_2(1) .$$

Then

(294) $$B = UV^*D_1$$

and

(295) $$V^*D_1 \in \mathcal{J}_2(1,w,w^*) .$$

By applying (289), (290), (291) with $l_1, l_2, \ldots, l_w$ instead of n we easily get the existence of a

(296) $$V \in \Delta_2(1^*) ,$$

such that

(297) $$V^*D_1 = DV$$

with (276) holds. From (294), (297) we obtain

(298) $$B = UDV$$

with (275), (276), (296) .

Hence each coset of $\mathcal{B}_2(1;w,w^*)/\Delta_2(1^*)$ possesses a representative B of type (274) with (275), (276).

The formula (277) is a trivial consequence of (274).

Finally we have to show that each coset of $\mathcal{B}_\alpha(1;w,w^*)/\Delta_\alpha(1^*)$ ($\alpha = 1,2$) possesses exactly one representative (274) with (275), (276). Suppose $\hat{B} = \hat{U}\hat{D}$ and $\check{B} = \check{U}\check{D}$ with

(299) $$\hat{U},\check{U} \in \check{\mathcal{M}}(1),$$

(300) $$\hat{D},\check{D} \in \mathcal{J}^*(1;w,w^*)$$

represent the same coset of $\mathcal{B}_\alpha(1;w,w^*)/\Delta_\alpha(1^*)$ ($\alpha = 1,2$). Then

(301) $$\hat{U}\hat{D} = \check{U}\check{D}V$$

with

(302) $$V \in \Delta_\alpha(1^*) \qquad (\alpha = 1,2) .$$

But $\check{\mathcal{M}}(1) \in \Psi_2(1)$ and $\mathcal{J}^*(1;w,w^*) \subset \mathcal{J}_2(1;w,w^*)$. Hence $\hat{U}\hat{D},\check{U}\check{D} \in \mathcal{B}_2(1;w,w^*)$ and from (301), (302) we get

(303) $$V \in \Delta_2(1^*) .$$

From (301) we get

(304) $$\check{U}^{-1}\hat{U}\hat{D} = \check{D}V .$$

Then (300), (303), (304) show us

(305) $$\check{U}^{-1}\hat{U} \in \Delta_2(1)$$

and because of (299) it is

$$(306)\qquad \hat{U} = \check{U} .$$

The formulas (301), (306) give

$$(307)\qquad \hat{D} = \check{D}V .$$

From (300), (307) we deduce $V = E$ and

$$(308)\qquad \hat{D} = \check{D} .$$

Herewith it is shown that the cosets $\mathcal{B}_\alpha(1;w,w^*)/\Delta_\alpha(1^*)$ $(\alpha = 1,2)$ have exactly one representative of the described form. Theorem 29 is proved.

THEOREM 30: The groups $\Psi'_\alpha(1)$, $\Delta'_\alpha(1)$ may be obtained by transposing all elements of $\Psi_\alpha(1)$, $\Delta_\alpha(1)$. Then

$$(309)\qquad W(n)\Psi'_\alpha(1)W(n) = \Psi_\alpha(\tilde{1}),\ W(n)\Delta'_\alpha(1)W(n) = \Delta_\alpha(\tilde{1}) \qquad (\alpha = 1,2) .$$

Let $U = (\overset{1}{U}_{\nu\mu}) \in \Psi_\alpha(1)$ and set

$$(310)\qquad \tilde{U} = (\overset{\tilde{1}}{\tilde{U}}_{\nu\mu}) = W(n)U'^{-1}W(n) \in \Psi_\alpha(\tilde{1}) \qquad (\alpha = 1,2) .$$

Then

$$(311)\qquad (\mathrm{Det}\ \overset{\tilde{1}}{\tilde{U}}_\nu)(\mathrm{Det}\ \overset{1}{U}_{w+1-\nu}) \equiv 1 \bmod q \qquad (\nu = 1,\dots,w) .$$

PROOF: Clear.

THEOREM 31: Let A_ν be integral $k_{\nu+1} \times k_\nu$ matrices $(\nu=1,\dots,w-1)$. Then

$$(312)\qquad B_{\nu\mu} = A_\nu A_{\nu-1} \cdots A_\mu \quad (1 \le \mu \le \nu \le w-1)$$

are integral $k_{\nu+1} \times k_\mu$ matrices. Furthermore the following divisibility properties hold

(313) $$\langle B_{\nu,\nu-1}\rangle \mid \langle B_{\nu\nu}\rangle\langle B_{\nu-1,\nu-1}\rangle \quad (\nu=2,\ldots,w-1),$$

(314) $$\langle B_{\nu-1,\mu}\rangle\, \langle B_{\nu,\mu-1}\rangle \mid \langle B_{\nu\mu}\rangle\langle B_{\nu-1,\mu-1}\rangle \quad (1<\mu<\nu<w),$$

(315) $$\langle B_{\nu-1,1}\rangle \mid \langle B_{\nu 1}\rangle \qquad (\nu = 2,\ldots,w-1).$$

If A_ν is of type

(316) $$A_\nu = \begin{pmatrix} \begin{matrix}1\\ A_{\nu 1}\end{matrix} & & & * \\ & \ddots & & \\ 0 & & \begin{matrix}1\\ A_{\nu\nu}\end{matrix} & \\ & & & \\ 0 & & 0 & \end{pmatrix} \qquad (\nu = 1,\ldots,w-1)$$

with $1_\mu \times 1_\mu$ matrices $A_{\nu\mu}$ $(\mu = 1,\ldots,\nu)$, then

(317) $$\langle B_{\nu,\nu-1}\rangle(\text{abs } A_{\nu\nu}) = \langle B_{\nu\nu}\rangle\langle B_{\nu-1,\nu-1}\rangle \quad (\nu = 2,\ldots,w-1)\ ,$$

(318) $$\langle B_{\nu-1,\mu}\rangle\langle B_{\nu,\mu-1}\rangle(\text{abs } A_{\nu\mu}) = \langle B_{\nu\mu}\rangle\langle B_{\nu-1,\mu-1}\rangle \quad (1<\mu<\nu<w),$$

(319) $$\langle B_{\nu-1,\mu}\rangle(\text{abs } A_{\nu 1}) = \langle B_{\nu 1}\rangle \qquad (\nu = 2,\ldots,w-1)\ .$$

PROOF: With

(320) $$U_\nu \in \Omega(k_{\nu+1}) \qquad (\nu = 1,\ldots,w-1)$$

set

(321) $$A_1^* = U_1\, A_1,\ A_\nu^* = U_\nu\, A_\nu\, U_{\nu-1}^{-1} \quad (\nu = 2,\ldots,w-1)\ .$$

Then

(322) $$B_{\nu\mu}^* = A_\nu^* \ldots A_\mu^* = U_\nu\, B_{\nu\mu}\, U_{\mu-1}^{-1} \quad (1 \le \mu \le \nu \le w-1)\ ,$$

and therefore

$$(323)\qquad \langle B^*_{\nu\mu}\rangle = \langle B_{\nu\mu}\rangle \qquad (1 \le \mu \le \nu \le w-1)\ .$$

By suitable choice of U_1 we first bring A_1^* into the form (316). If $U_{\nu-1}$ is already fixed we choose U_ν such that A_ν^* obtains the form (316). Hence we may assume that all A_ν^* are of type (316). Hence it suffices to prove (313), (314), (315) for matrices of type (316). But in this case (313), (314), (315) follow from (317), (318), (319). Hence it suffices to prove the latter formulas.

From (312), (316) we get

$$(324)\qquad B_{\nu\mu} = \begin{pmatrix} B_{\nu\mu 1} & & * \\ & \ddots & \\ & & B_{\nu\mu\mu} \\ 0 & & 0 \end{pmatrix} \qquad (1 \le \mu \le \nu \le w-1)$$

with

$$(325)\qquad B_{\nu\mu\iota} = A_{\nu\iota} A_{\nu-1,\iota} \cdots A_{\mu\iota} \quad (1 \le \iota \le \mu \le \nu \le w-1).$$

It follwos

$$(326)\qquad \langle B_{\nu\mu}\rangle = \prod_{\iota=1}^{\mu} \operatorname{abs} B_{\nu\mu\iota} = \prod_{\lambda=\mu}^{\nu} \prod_{\iota=1}^{\mu} \operatorname{abs} A_{\lambda\iota} \quad (1 \le \mu \le \nu \le w-1).$$

From this (317), (318), (319) follow immediately. Theorem 31 is proved.

Put

$$(327)\qquad Q(1) = \begin{pmatrix} W(n^*) & 0 \\ 0 & qE^{(1_w)} \end{pmatrix} ,$$

$$(328)\qquad P(1) = q^{\frac{1_w}{n}} W(n)Q(1)^{-1} = \begin{pmatrix} 0 & q^{\frac{1_w}{n}-1} W(1_w) \\ q^{\frac{1_w}{n}} E^{(n^*)} & 0 \end{pmatrix} ,$$

$$(329)\qquad \mathring{F}(l) = \begin{pmatrix} W(l_1) & & & 0 \\ & \cdot & & \\ & & \cdot & \\ 0 & & \cdot & W(l_w) \end{pmatrix} \in \Delta_1(l) .$$

Then

$$(330)\qquad \operatorname{abs} P(l) = 1 ,$$

$$(331)\qquad P(\mathfrak{z}^{w-1}l)P(\mathfrak{z}^{w-2}l) \ldots P(l) = \mathring{F}(l) ,$$

$$(332)\qquad W(n)P'(l)W(n) = P(l) .$$

THEOREM 32: It is

$$(333)\qquad P(l)\Psi_1(l)P^{-1}(l) = \Psi_1(\mathfrak{z}l) .$$

Let

$$(334)\qquad U = (\overset{l}{U}_{\nu\mu}) \in \Psi_1(l)$$

and

$$(335)\qquad U^* = (\overset{\mathfrak{z}l}{U}{}^*_{\nu\mu}) = P(l)UP^{-1}(l) .$$

Then

$$(336)\qquad \overset{\mathfrak{z}l}{U}{}^*_1 = W(l_w)\, \overset{l}{U}_w\, W(l_w), \quad \overset{\mathfrak{z}l}{U}{}^*_\nu = \overset{l}{U}_{\nu-1} \quad (\nu = 2, \ldots, w).$$

PROOF: Easy computation

THEOREM 33: Set

$$(337)\qquad r(l) = q^{l_w(l_1+\ldots+l_{w-2})} ,$$

let all elements of the $l_w \times (l_1+\ldots+l_{w-2})$ matrix L run over the numbers $0,1,\ldots,q-1$ and form the $r(l)$ matrices

$$(338)\qquad K_\rho(l) = \begin{pmatrix} P^{-1}(l^*) & 0 \\ 0 & q^{1-\frac{l_{w*}}{n^*}} E^{(l_w)} \end{pmatrix} \begin{pmatrix} E^{(l_{w*})} & 0 & 0 \\ 0 & E^{(l_1+\ldots+l_{w-2})} & 0 \\ 0 & L & E^{(l_w)} \end{pmatrix}.$$

Then

$$(339)\qquad \operatorname{abs} K_\rho(l) = q^{\frac{l_w k_{w-2}}{n^*}} ,$$

$$(340)\qquad \mathcal{L}_1(l;w,w^*)P^{-1}(l^*) = \bigcup_{\rho=1}^{r} K_\rho(l)\mathcal{L}_1(\hat{l};w,w^*) ,$$

with

$$(341)\qquad K_\rho\mathcal{L}_1(\hat{l};w,w^*) \cap K_\lambda\mathcal{L}_1(\hat{l};w,w^*) = \emptyset \quad (\rho \neq \lambda;\ \lambda,\rho = 1,\ldots,r) .$$

Furthermore let

$$(342)\qquad A = (\overset{l}{A}_{\nu\mu}) \in \mathcal{L}_1(l;w,w^*) ,$$

$$(343)\qquad A^* = (\overset{\hat{l}_*}{A^*_{\nu\mu}}) \in \mathcal{L}_1(\hat{l};w,w^*)$$

and

$$(344)\qquad AP^{-1}(l^*) = K_\rho A^* \quad \text{for certain } \rho \quad (1 \leq \rho \leq r).$$

Then

$$(345)\qquad \overset{\hat{l}_*}{A^*_1} = W(l_{w*})\overset{l}{A}_{w*}W(l_{w*}),\ \overset{\hat{l}_*}{A^*_\nu} = \overset{l}{A}_{\nu-1} \qquad (\nu = 2,\ldots,w^*) .$$

PROOF: An elementary computation shows (345),

(346) $$\overset{\wedge}{1}_* A^*_{\iota\varkappa} \equiv 0 \bmod q \qquad (1 \leq \varkappa < \iota \leq w^*) ,$$

(347) $$\overset{\wedge}{1}_* A^*_{w1} \equiv 0 \bmod q ,$$

(348) $$(A^*_{w2} \ldots A^*_{ww^*}) = LB + q^{-1}(A_{w1}, \ldots, A_{w,w-2})$$

with

(349) $$B = \begin{pmatrix} A_{11} & \cdots & A_{1,w-2} \\ \vdots & & \vdots \\ A_{w-2,1} & \cdots & A_{w-2,w-2} \end{pmatrix} .$$

But

(350) $$\langle \mathrm{Det}\, B, q \rangle = 1 .$$

Hence there is exactly one L with

(351) $$LB + q^{-1}(A_{w1}, \ldots, A_{w,w-2}) \equiv 0 \bmod q,$$

that means

(352) $$A^*_{w\nu} \equiv 0 \bmod q \qquad (\nu = 2, \ldots, w^*) .$$

From this theorem 33 follows.

§ 6. THE RIEMANNIAN SPACE OF POSITIVE MATRICES

The space of all real positive $n \times n$ matrices Y forms a "weakly symmetric Riemannian space" in the sense of Selberg [40]. We prove some basic results on this space which are about identical with the results in Maaß [33], § 6.

The space $\mathcal{Y}(n)$ of all real symmetric $n \times n$ matrices $Y > 0$ has the real dimension

$$(353) \qquad d(n) = \frac{n(n+1)}{2} \; .$$

By the substitution

$$(354) \qquad Y \to Y[A]$$

it is mapped onto itself. Here A is a non-singular real $n \times n$ matrix. Let $[dY]$ denote the Euclidean volumelement in $\mathcal{Y}(n)$. Then

$$(355) \qquad dv_Y = (\mathrm{Det}\ Y)^{-\frac{n+1}{2}} [dY]$$

is invariant under (354) and under

$$(356) \qquad Y \to Y^{-1}.$$

The proof can be found in Maaß [33], § 6. Furthermore the proof of the following results stands in Christian [7], IV.1.

Each matrix $Y \in \mathcal{Y}(n)$ may be uniquely written as

$$(357) \qquad Y = R[D] \; .$$

Here

$$(358) \qquad R = [r_1, \ldots, r_n]$$

is a diagonalmatrix and

$$(359) \qquad D = \begin{pmatrix} 1 & & & d_{1\varkappa} \\ & \cdot & & \\ & & \cdot & \\ 0 & & \cdot & 1 \end{pmatrix}$$

an upper unipotent matrix. One has

$$(360) \qquad r_1 = y_1$$

(361) $$0 < r_\nu \leq y_\nu \qquad (\nu = 1,\dots,n) .$$

The relation (357) is called "Jacobi's transformation".

For $\mu > 0$ the "Siegeldomain" $\mathcal{S}(n,\mu) \subset \mathcal{Y}(n)$ is defined by

(362) $$\frac{r_1}{r_2}, \frac{r_2}{r_3}, \dots, \frac{r_{n-1}}{r_n} < \mu ,$$

(363) $$-\mu < d_{\iota\varkappa} < \mu \qquad (1 \leq \iota < \varkappa \leq n).$$

There exists a constant $c_{10} = c_{10}(n,\mu) > 1$ such that for

(364) $$Y \in \mathcal{S}(n,\mu)$$

we have

(365) $$1 \leq \frac{y_\nu}{r_\nu} \leq c_{10} \qquad (\nu = 1,\dots,n) ,$$

(366) $$c_{10}^{-1} \operatorname{Dg} Y \leq Y \leq c_{10} \operatorname{Dg} Y ,$$

(367) $$\frac{y_1}{y_2}, \frac{y_2}{y_3}, \dots, \frac{y_{n-1}}{y_n} < c_{10} .$$

By

(368) $$Y \to Y[U] \qquad (U \in \Omega(n))$$

the group $\Omega(n)$ operates discontinuously on $\mathcal{Y}(n)$. A fundamental domain is given by "Minkowski's pyramid" $\mathcal{M}(n)$. This is a convex pyramid with the cusp in the origin which is bounded by finitely many hyperplanes of $(d(n)-1)$ real dimensions. There exists a constant $c_{11} = c_{11}(n) > 0$ with

(369) $$\mathcal{M}(n) \subset \mathcal{S}(n,c_{11}) .$$

Therefore the inequalities (362), (363), (365) till (367) are especially true in $\mathcal{M}(n)$.

Let

$$(370)\qquad h_\alpha = h_\alpha(1) = [\Omega(n) : \Psi_\alpha(1)] \qquad (\alpha = 1,2)$$

and

$$(371)\qquad F_{\alpha 1},\dots, F_{\alpha h_\alpha} \in \Omega(n) \qquad (\alpha = 1,2)$$

with

$$(372)\qquad \Omega(n) = \bigcup_{\nu=1}^{h_\alpha} F_{\alpha\nu}\,\Psi_\alpha(1) \qquad (\alpha = 1,2)\ .$$

Then

$$(373)\qquad \mathfrak{F}_\alpha(1) = \bigcup_{\nu=1}^{h_\alpha} \mathfrak{M}(n)[F_{\alpha\nu}] \qquad (\alpha = 1,2)$$

is a fundamental domain of $\Psi_\alpha(1)$. Put

$$(374)\qquad \mathfrak{M}_1(n) = \{Y \in \mathfrak{M}(n);\ \mathrm{Det}\ Y \geq 1\}\ ,$$

$$(375)\qquad \mathfrak{F}_{\alpha 1}(1) = \{Y \in \mathfrak{F}_\alpha(1); \mathrm{Det}\ Y \geq 1\} \qquad (\alpha = 1,2),$$

$$(376)\qquad \mathfrak{F}_{\alpha 1}(1) = \bigcup_{\nu=1}^{h_\alpha} \mathfrak{M}_1(n)[F_{\alpha\nu}] \qquad (\alpha = 1,2)\ .$$

THEOREM 34: Let $\Phi(y)$ denote a complex-valued function which is continuous in $0 < y < \infty$. Then for $0 < y_1 < y_2$ one has

$$(377)\qquad \int\limits_{\substack{\mathfrak{M}(n)\\ y_1 < \mathrm{Det}\ Y < y_2}} \Phi(\mathrm{Det}\ Y)dv_Y = \frac{n+1}{2}\, v_n \int_{y_1}^{y_2} \Phi(y)\frac{dy}{y}$$

with

$$(378)\qquad v_n = \int\limits_{\substack{\mathfrak{M}(n)\\ \mathrm{Det}\ Y \leq 1}} [dY] < \infty\ .$$

PROOF: See Maaβ [33], page 145, Lemma 2.

THEOREM 35: Consider the generalized Jacobi's transformation

$$Y = R[D] , \tag{379}$$

$$R = \begin{pmatrix} R_1 & & & 0 \\ & R_2 & & \\ & & \ddots & \\ 0 & & & R_w \end{pmatrix} , \tag{380}$$

$$D = \begin{pmatrix} E & D_{12} & \cdots & D_{1w} \\ 0 & E & \cdots & D_{2w} \\ \vdots & & \ddots & \vdots \\ 0 & \cdots & & E \end{pmatrix} \tag{381}$$

with $l_\nu \times l_\nu$ matrices R_ν $(\nu = 1,\ldots,w)$ and $l_\nu \times l_\mu$ matrices $D_{\nu\mu}$ $(1 \leq \nu < \mu \leq w)$. Set

$$[dR] = \prod_{\nu=1}^{w} [dR_\nu] , \tag{382}$$

$$[dD] = \prod_{1 \leq \nu < \mu \leq w} [d\, D_{\nu\mu}] . \tag{383}$$

Then

$$[dY] = \prod_{\nu=1}^{w-1} (\text{Det } R_\nu)^{n-k_\nu} [dR][dD]. \tag{384}$$

PROOF: For $w = 2$ the theorem was proved in Christian [7], pages 193, 194. From this the theorem follows by induction with respect to w.

From (355), (384) we obtain

(385) $$dv_Y = \prod_{\nu=1}^{w} \{ (\mathrm{Det}\ R_\nu)^{\frac{1}{2}(n-k_\nu-k_{\nu-1})} dv_{R_\nu} \} [dD] .$$

For $Y \in \mathcal{Y}(n)$ set

(386) $$\tilde{Y} = (Y[W(n)])^{-1} = Y^{-1}[W(n)] ,$$

(387) $$\check{Y} = \overset{1}{\check{Y}} = (Y[Q(1)])^{-1} = Y^{-1}[Q(1)^{-1}] .$$

Then

(388) $$\tilde{\tilde{Y}} = \check{\check{Y}} = Y ,$$

(389) $$\mathrm{Det}\ \tilde{Y} = (\mathrm{Det}\ Y)^{-1},$$

(390) $$\mathrm{Det}\ \check{Y} = q^{-2l_w}(\mathrm{Det}\ Y)^{-1} ,$$

(391) $$\mathrm{Det}\ Y = q^{-\frac{2k_{w-2}l_w}{n^*}}\ \mathrm{Det}\ Y[K_\rho(1)] .$$

I shall now give a record about results which can in detail be found in Maaβ [33], §§ 5,6. One also may see in Maaβ [28], [29] and in Selberg [40].

Let dY be the differential of Y. By

(392) $$ds^2 = \mathrm{Tr}(Y^{-1}dY)^2$$

a positive definite metric is defined on $\mathcal{Y}(n)$ which is invariant under the substitutions (354) and (356). By this metric $\mathcal{Y}(n)$ becomes a "weakly symmetric Riemannian space" in the sense of Maaβ [33], § 5 and Selberg [40]. The differential operators, which are invariant under (354) form a commutative ring. Set

(393) $$e_{\mu\nu} = \begin{cases} 1 & \mu = \nu \\ \frac{1}{2} & \mu \neq \nu \end{cases} \qquad (\mu,\nu = 1,\ldots,n) ,$$

(394) $$\frac{\partial}{\partial Y} = \left(e_{\mu\nu} \frac{\partial}{\partial y_{\mu\nu}}\right) .$$

THEOREM 36: The differential operators

(395) $$\mathrm{Tr}\left(\left(Y \frac{\partial}{\partial Y}\right)^{\nu}\right) \qquad (\nu = 1,\dots,n)$$

form an algebraically independent basis of the commutative ring of invariant differential operators.

PROOF: See Maaβ [33], page 64, theorem.

Each invariant differential operator $L(Y)$ possesses with respect to the metric (392) exactly one adjoint $\hat{L}$ and

(396) $$\hat{L}(Y) = \bar{L}(Y^{-1}) ,$$

where $^{-}$ means the conjugate complex. Furthermore

(397) $$\hat{\hat{L}} = L .$$

For the proof see Maaβ [33], pages 58 till 60, page 68 formula

$$\widetilde{L\left(Y \frac{\partial}{\partial Y}\right)} = L\left(\hat{Y} \frac{\partial}{\partial \hat{Y}}\right) , \quad \hat{Y} = Y^{-1}$$

and page 78 formula $\hat{L} = \tilde{\bar{L}}$. The adjoint has the property

(398) $$\int_{\mathcal{Y}(n)} f(\hat{L}g)dv_Y = \int_{\mathcal{Y}(n)} (Lf)g \, dv_Y$$

provided that $(L_\mu^* f)(L_\nu^* g)$ vanish on the boundary of $\mathcal{Y}(n)$. Here L_ν^* are finitely many differential operators derived from L. (398) is the generalization of partial integration.

A special invariant differential operator is

(399) $$M_n = (\mathrm{Det}\, Y)\left(\mathrm{Det} \frac{\partial}{\partial Y}\right) .$$

See Maaß [33], page 67. For $k \in \mathbb{R}$ we form the invariant differential operator

(400) $$P_k = (\text{Det } Y)^{-k} \hat{M}_n (\text{Det } Y)^k M_n .$$

Then

(401) $$\hat{P}_k = (\text{Det } Y)^k M_n (\text{Det } Y)^{-k} \hat{M}_n = \hat{M}_n (\text{Det } Y)^k M_n (\text{Det } Y)^{-k} ,$$

(402) $$\hat{P}_k = (\text{Det } Y)^k P_k (\text{Det } Y)^{-k} .$$

See Maaß [33], page 210 .

Put

(403) $$D^*(k) = D^*(k,Y) = (\text{Det } Y)^{\frac{k}{4}} P_{\frac{k}{2}} (\text{Det } Y)^{-\frac{k}{4}} .$$

Since $D^*(k)$ is a real operator we obtain from (396), (402), (403)

(404) $$\hat{D}^*(k) = D^*(k) ,$$

i. e., $D^*(k)$ is self-adjoint. Furthermore $D^*(k)$ is an invariant operator. Hence from (386), (387) we obtain

(405) $$D^*(k,Y) = D^*(k,Y^{-1}) = D^*(k,\tilde{Y}) = D^*(k,\check{Y}) .$$

This operator is the generalization of $-D^*(t)$ defined in (19).

With the help of the metric (392) a distance $\rho(Y,Y^*)$ can be defined between any two points $Y,Y^* \in \mathcal{Y}(n)$. As before let Y_ν be the left upper $\nu\times\nu$ submatrix of Y.

THEOREM 37 It is

(406) $$\rho(Y_\nu,Y_\nu^*) \leq \rho(Y,Y^*) \qquad (\nu = 1,\dots,n) .$$

PROOF: See Maaß [33], page 145, Lemma 1.

Let

$$(407) \qquad s = (s_1,\ldots,s_w)$$

be a row of complex variables. Set

$$(408) \qquad \sigma_\nu = \mathrm{Re}\, s_\nu \qquad (\nu=1,\ldots,w)\ ,$$

$$(409) \qquad \sigma = (\sigma_1,\ldots,\sigma_w) = \mathrm{Re}\, s\ ,$$

$$(410) \qquad s^* = (s_1,\ldots,s_{w^*})\ ,$$

$$(411) \qquad \tilde{s}_\nu = -s_{w+1-\nu} \qquad (\nu = 1,\ldots,w)\ ,$$

$$(412) \qquad \tilde{s} = (\tilde{s}_1,\ldots,\tilde{s}_w)\ ,$$

$$(413) \qquad \check{s}_\nu = -s_{w-\nu} \quad (\nu = 1,\ldots,w-1),\ \check{s}_w = -s_w\ ,$$

$$(414) \qquad \check{s} = (\check{s}_1,\ldots,\check{s}_w)\ .$$

Then

$$(415) \qquad \check{s} = (\widetilde{s^*},\ -s_w)\ ,$$

$$(416) \qquad \tilde{\tilde{s}} = \check{\check{s}} = s\ .$$

Put

$$(417) \qquad \mathfrak{z}s = (s_w,s_1,\ldots,s_{w-1})\ ,$$

$$(418) \qquad \mathfrak{z}^*s^* = (s_{w-1},s_1,\ldots,s_{w-2})\ ,$$

$$(419) \qquad \hat{s}_1 = s_{w^*},\ \hat{s}_\nu = s_{\nu-1}\ (\nu = 2,\ldots,w^*),\hat{s}_w = s_w,$$

$$(420) \qquad \hat{s} = (\hat{s}_1,\ldots,\hat{s}_w)\ ,$$

$$(421) \qquad \hat{s} = (\mathfrak{z}^*s^*,s_w)\ .$$

Set

(422) $$\{l,s\} = \frac{1}{n}\sum_{\nu=1}^{w} l_\nu s_\nu .$$

Then

(423) $$\{l,s\} = \{\mathfrak{z}l,\mathfrak{z}s\} = \{\hat{l},\hat{s}\} = -\{\tilde{l},\tilde{s}\} = -\{\check{l},\check{s}\}.$$

THEOREM 38: The function

(424) $$f(l,Y,s) = (\operatorname{Det} Y)^{s_w + \frac{n^*}{4}} \prod_{\nu=1}^{w-1} (\operatorname{Det} Y_{k_\nu})^{s_\nu - s_{\nu+1} - \frac{l_{\nu+1}+l_\nu}{4}}$$

is homogeneous in Y of degree $n\{l,s\}$, and it is an eigenfunction of all invariant differential operators. Let

(425) $$e(w) = (1,\ldots,1)$$

with elements 1 and $a \in \mathbb{C}$. Then

(426) $$(\operatorname{Det} Y)^a f(l,Y,s) = f(l,Y,ae(w) + s) ,$$

(427) $$f(l,Y,s) = (\operatorname{Det} Y)^{s_w + \frac{n^*}{4}} f(l^*,Y_{n^*},s^* - (s_w + \tfrac{n}{4})e(w^*)) .$$

Let $D = (\overset{l}{D}_{\nu\mu})$ be a real matrix with $\overset{l}{D}_{\nu\mu} = 0$ $(1 \leq \mu < \nu \leq w)$.
Then

(428) $$f(l,Y[D],s) = \left(\prod_{\mu=1}^{w} (\operatorname{abs} D_\mu)^{2s_\mu + \frac{1}{2}(k_\mu + k_{\mu-1} - n)}\right) f(l,Y,s).$$

Especially

(429) $$f(l,Y[V],s) = f(l,Y,s) \quad (V \in \Delta_\alpha(l);\ \alpha = 1,2) .$$

Let $\omega(\iota)$ be given by definition 1 and set

(430) $$p_\iota = s_{\omega(\iota)} + \frac{1 + k_{\omega(\iota)} + k_{\omega(\iota)-1}}{4} - \frac{1}{2} \quad (\iota = 1,\ldots,n),$$

(431) $$p = (p_1,\ldots,p_n)\ .$$

Then

(432) $$f(l,Y,s) = f(\overset{\circ}{l},Y,p)\ .$$

PROOF: The formulas follow from an easy computation. That $f(l,Y,s)$ is an eigenfunction of all invariant differential operators follows from (432) and Maaβ [33], page 69.

THEOREM 39: It is

(433) $$f(\tilde{l},\tilde{Y},\tilde{s}) = f(l,Y,s)\ .$$

PROOF: From the theory of determinates we get the equation

(434) $$\mathrm{Det}(Y_\nu) = (\mathrm{Det}\ Y)\mathrm{Det}((\tilde{Y})_{n-\nu}) \qquad (\nu = 1,\ldots,n)\ .$$

From (223), (227), (411), (424), (434) we get

$$f(l,Y,s) = (\mathrm{Det}\ Y)^{s_w+\frac{k_{w-1}}{4}+\sum_{\nu=1}^{w-1}\left(s_\nu-s_{\nu+1}-\frac{l_{\nu+1}+l_\nu}{4}\right)} \times$$

$$\prod_{\nu=1}^{w-1}(\mathrm{Det}(\tilde{Y})_{n-k_\nu})^{s_\nu-s_{\nu+1}-\frac{l_{\nu+1}+l_\nu}{4}} =$$

$$(\mathrm{Det}\ Y)^{-\tilde{s}_w-\frac{\tilde{k}_{w-1}}{4}} \prod_{\nu=1}^{w-1}(\mathrm{Det}(\tilde{Y})_{\tilde{k}_{w-\nu}})^{\tilde{s}_{w-\nu}-\tilde{s}_{w-\nu+1}-\frac{\tilde{l}_{w-\nu+1}+\tilde{l}_{w-\nu}}{4}}$$

$$= (\mathrm{Det}\ \tilde{Y})^{\tilde{s}_w+\frac{\tilde{k}_{w-1}}{4}} \prod_{\nu=1}^{w-1}(\mathrm{Det}(\tilde{Y})_{\tilde{k}_\nu})^{\tilde{s}_\nu-\tilde{s}_{\nu+1}-\frac{\tilde{l}_{\nu+1}+\tilde{l}_\nu}{4}} =$$

$$f(\tilde{l},\tilde{Y},\tilde{s})\ .$$

Theorem 39 is proved.

THEOREM 40: Let $X \in \eta(n)$ and put

(435) $$J(l,X,s) = \int_{Y \in \eta(n)} f(l,Y,s)\exp(-\mathrm{Tr}(X^{-1}Y))dv_Y .$$

Then

(436) $$J(l,X,s) = \pi^{\frac{n(n-1)}{4}} \Big(\prod_{\nu=1}^{w} \prod_{\iota=0}^{l_\nu-1} \Gamma\big(s_\nu + \frac{l_\nu - n}{4} - \frac{\iota}{2}\big)\Big) f(l,X,s) .$$

PROOF: Because of (432) it suffices to consider the case $l = \mathring{l}$ and $w = n$. Then we have to prove

(437) $$J(\mathring{l},X,s) = \pi^{\frac{n(n-1)}{4}} \prod_{\nu=1}^{n} \Gamma\big(s_\nu - \frac{n-1}{4}\big) \cdot f(\mathring{l},X,s) .$$

Put

(438) $$X = T'T$$

with an upper triangular matrix

(439) $$T = \begin{pmatrix} t_1 & t_{12} & \cdots & t_{1n} \\ & \ddots & & \vdots \\ 0 & & & t_n \end{pmatrix} .$$

Then

(440) $$\mathrm{Det}(X_\nu) = t_1^2 \cdots t_\nu^2 \qquad (\nu = 1,\dots,n) .$$

The substitution $Y \to Y[T]$ gives

(441) $$J(\mathring{l},X,s) = \int_{Y \in \eta(n)} f(\mathring{l},Y[T],s)\exp(-\mathrm{Tr}\, Y)dv_Y .$$

From (428) we get

$$(442)\qquad f(\mathring{1},Y[T],s) = \Big(\prod_{\mu=1}^{n} t_\mu^{2s_\mu+\mu-\frac{n+1}{2}}\Big) f(\mathring{1},Y,s) .$$

Applying (440) gives us

$$(443)\qquad \prod_{\mu=1}^{n} t_\mu^{2s_\mu+\mu-\frac{n+1}{2}} = f(\mathring{1},X,s) .$$

Hence

$$(444)\qquad f(\mathring{1},Y[T],s) = f(\mathring{1},X,s)f(\mathring{1},Y,s) .$$

Inserting this into (441) gives

$$(445)\qquad J(\mathring{1},X,s) = J(\mathring{1},E,s)f(\mathring{1},X,s) .$$

Therefore it suffices to prove

$$(446)\qquad J(\mathring{1},E,s) = \pi^{\frac{n(n-1)}{4}} \prod_{\nu=1}^{n} \Gamma\Big(s_\nu - \frac{n-1}{4}\Big) .$$

Perform the Jacobi transformation (379), i. e. $Y = R[D]$. Because this is similar to (438), formula (443) gives

$$(447)\qquad f(\mathring{1},Y,s) = \prod_{\mu=1}^{n} r_\mu^{s_\mu+\frac{\mu}{2}-\frac{n+1}{4}} .$$

Furthermore

$$(448)\qquad \operatorname{Tr} Y = \sum_{\mu=1}^{n} r_\mu + \sum_{1\le\mu<\nu\le n} r_\mu d_{\mu\nu}^2$$

and from (385)

$$(449)\qquad dv_Y = \prod_{\mu=1}^{n}\Big(r_\mu^{\frac{n+1}{2}-\mu} dv_{r_\mu}\Big)[dD] .$$

Hence

$$(450)\quad J(\mathring{1},E,s) = \int\limits_{r_1,\dots,r_n>0} \left(\prod_{\mu=1}^{n} r_\mu^{s_\mu - \frac{\mu}{2} + \frac{n+1}{2}}\right) \times$$

$$\exp\left(-\sum_{\mu=1}^{n} r_\mu - \sum_{1\le\mu<\nu\le n} r_\mu d_{\mu\nu}^2\right) dv_{r_1}\dots dv_{r_n}[dD]$$

Performing the substitution $u_{\mu\nu} = \sqrt{r_\mu}\, d_{\mu\nu}$ gives

$$(451)\quad J(\mathring{1},E,s)=\left(\int_{-\infty}^{\infty}\exp(-u^2)du\right)^{\frac{n(n-1)}{2}} \left(\prod_{\mu=1}^{n}\int_0^\infty r_\mu^{s_\mu - \frac{n-1}{4}} \exp(-r_\mu)dv_{r_\mu}\right).$$

This gives (446). Theorem 40 is proved.

Perform the variable transformation

$$(452)\qquad u_\nu = s_{\nu+1} - s_\nu + \frac{l_{\nu+1} + l_\nu}{4} \qquad (\nu = 1,\dots,w-1),$$

$$(453)\qquad u = (u_1,\dots,u_{w-1}).$$

Set

$$(454)\qquad \hat{f}(l,Y,u,s_w + \frac{n^*}{4}) = f(l,Y,s).$$

Hence

$$(455)\qquad \hat{f}(l,Y,u,a) = (\operatorname{Det} Y)^a \prod_{\nu=1}^{w-1} (\operatorname{Det} Y_{k_\nu})^{-u_\nu}$$

§ 7 THETA FUNCTIONS

In this paragraph we deduce results on thetafunctions which will be needed in the next chapter in order to prove analytic continuation and functional equations of Selberg's zetafunctions and L-series.

THEOREM 41: Let

(456) $$Y \in \mathfrak{Y}(n),\ T \in \mathfrak{Y}(m)$$

and U,V complex $n \times n$ matrices. Then

(457) $$\sum_A \exp(-\pi\ \mathrm{Tr}(Y[A+V]T + 2\pi i\ A'U)) =$$
$$\exp(-2\pi i\ \mathrm{Tr}(U'V))(\mathrm{Det}\ Y)^{-\frac{m}{2}}(\mathrm{Det}\ T)^{-\frac{n}{2}} \times$$
$$\sum_A \exp(-\pi\ \mathrm{Tr}(Y^{-1}[A-U]T^{-1} + 2\pi i\ A'V)) .$$

A runs over all integral $n \times m$ matrices.

PROOF: For a $n \times n$ matrix $C = (c_{\iota\varkappa})$ and a $m \times m$ matrix $B = (b_{\iota\varkappa})$ form the $mn \times mn$ matrix

(458) $$\mathfrak{Y}(C,B) = \begin{pmatrix} b_{11}C & b_{12}C & \cdots & b_{1m}C \\ b_{21}C & b_{22}C & \cdots & b_{2m}C \\ \vdots & \vdots & & \vdots \\ b_{m1}C & b_{m2}C & \cdots & b_{mm}C \end{pmatrix}$$

If one takes all elements of C and B as indeterminates one has decompositions

(459) $$C = C_1C_2\ ,\ B = B_1B_2$$

with

(460) $$\left\{\begin{array}{ll} C_1 = \begin{pmatrix} * & & 0 \\ \vdots & \ddots & \\ * & \cdots & * \end{pmatrix} & ,\ C_2 = \begin{pmatrix} * & & * \\ & \ddots & \\ 0 & & * \end{pmatrix}, \\ B_1 = \begin{pmatrix} * & & 0 \\ \vdots & \ddots & \\ * & \cdots & * \end{pmatrix} & ,\ B_2 = \begin{pmatrix} * & & * \\ & \ddots & \\ 0 & & * \end{pmatrix} \end{array}\right\} .$$

Then $\mathfrak{F}(C_1,B_1)$ is a lower triangular matrix and $\mathfrak{F}(C_2,B_2)$ an upper triangular matrix and one has

(461) $$\mathfrak{F}(C,B) = \mathfrak{F}(C_1,B_1)\mathfrak{F}(C_2,B_2) .$$

From this one easily sees

(462) $$\operatorname{Det} \mathfrak{F}(C,B) = (\operatorname{Det} C)^m(\operatorname{Det} B)^n ,$$

(463) $$(\mathfrak{F}(C,B))^{-1} = \mathfrak{F}(C^{-1},B^{-1}) .$$

Since this holds for indeterminate C,B, the formula (462) holds for all complex C,B; formula (463) holds for complex non-singular C,B. Let C,B be real symmetric and positive. Then one may take $C_1 = C_2'$, $B_1 = B_2'$. Then (461) becomes

(464) $$\mathfrak{F}(C,B) = \mathfrak{F}(C_2,B_2)'\mathfrak{F}(C_2,B_2) .$$

Hence $\mathfrak{F}(C,B)$ is symmetric and positive.

Now decompose A,U,V in n-rowed colums

(465) $$A = (a_1,\dots,a_m);\ U = (u_1,\dots,u_m);\ V = (v_1,\dots,v_m)$$

and form the mn-rowed columns

(466) $$a = \begin{pmatrix} a_1 \\ \vdots \\ a_m \end{pmatrix},\quad u = \begin{pmatrix} u_1 \\ \vdots \\ u_m \end{pmatrix},\quad v = \begin{pmatrix} v_1 \\ \vdots \\ v_m \end{pmatrix} .$$

Then (457) is identical with

(467) $$\sum_a \exp(-\pi(\mathfrak{F}(Y,T))[a+v] + 2\pi i\ a'u) =$$

$$\exp(-2\pi i u'v)(\operatorname{Det} \mathfrak{F}(Y,T))^{-\frac{1}{2}} \sum_a \exp(-\pi(\mathfrak{F}^{-1}(Y,T))[a-u] + 2\pi i a'v) .$$

But this follows from (3). Theorem 41 is proved.

DEFINITION 2: For $m \in \mathbb{N}$ and an even character χ mod q form the Gaussian sum

(468) $$G(m,\chi,C) = q^{-\frac{m^2}{2}} \sum_{D \bmod q} \chi(\operatorname{Det} D)\exp(\frac{2\pi i}{q} \operatorname{Tr}(C'D)) ,$$

where D runs over all integral $m \times m$ matrices mod q and set

(469) $$G(m,\chi) = G(m,\chi,E) .$$

THEOREM 42: Let χ be an even primitive character mod q. Then

(470) $$G(m,\chi,C) = \bar{\chi}(\operatorname{Det} C)G(m,\chi)$$

for all integral $m \times m$ matrices C.

Furthermore

(471) $$\operatorname{abs} G(m,\chi) = 1 .$$

PROOF: Formula (470) is mentioned without proof already in Andrianov [2], page 41. For $m = 1$ the theorem is proved in Landau [22], pages 484 - 485. The following proof is a generalization of Landau's proof.

First let

(472) $$\langle \operatorname{Det} C,q\rangle = 1 .$$

Then with D also $D^* = C'D$ runs over all residue classes mod q. Hence from (468)

$$G(m,\chi,C) = \bar{\chi}(\operatorname{Det} C)q^{-\frac{m^2}{2}} \sum_{D \bmod q} \chi(\operatorname{Det}(C'D))\exp(\frac{2\pi i}{q} \operatorname{Tr}(C'D)) =$$

$$\bar{\chi}(\operatorname{Det} C)q^{-\frac{m^2}{2}} \sum_{D^* \bmod q} \chi(\operatorname{Det} D^*)\exp(\frac{2\pi i}{q} \operatorname{Tr} D^*) =$$

$$\bar{\chi}(\mathrm{Det}\ C)G(m,\chi) .$$

Now let

(473) $$\langle \mathrm{Det}\ C,q\rangle > 1 .$$

We have to prove

(474) $$G(m,\chi,C) = 0 .$$

Let $U,V \in \Omega(m)$. Then with D also $U'DV'$ runs over all residue classes mod q. Hence

(475) $$G(m,\chi,UCV) = G(m,\chi,C) \quad (U,V \in \Omega(m)) .$$

Therefore by the theorem of elementary divisors we may take

(476) $$C = [c_1,\ldots,c_m]$$

with

(477) $$c_1,\ldots,c_m \geq 0$$

and

(478) $$c_1 | c_2 | \ \ldots \ | \ c_m .$$

Then from (473), (478) we deduce

(479) $$\langle c_m,q\rangle = t > 1 .$$

Set

(480) $$c_m = tb, \ q = tr,$$

(481) $$D = \begin{pmatrix} D_1^{(m-1,m)} \\ d \end{pmatrix} ,$$

(482) $$d = kr + n$$

with

(483) $$k = (k_1,\dots,k_m)\ ,\ n = (n_1,\dots,n_m)\ .$$

Then

(484) $$G(m,\chi,C) =$$

$$q^{-\frac{m^2}{2}} \sum_{\substack{D_1 \bmod q \\ n \bmod r \\ k \bmod t}} \chi(\mathrm{Det}\ D)\exp\left(2\pi i\ \frac{c_1d_1+\dots+c_{m-1}d_{m-1}}{q} + \frac{bn_m}{r}\right)\ .$$

We shall show, that there exists an integer a with

(485) $$\langle a,q\rangle = 1\ ,$$

(486) $$a \equiv 1 \bmod r\ ,$$

(487) $$\chi(a) \neq 1\ .$$

Let $\mathfrak{m}$ be the set of integers which satisfy (485), (486). Because of $1 \in \mathfrak{m}$ it follows that $\mathfrak{m} \neq \emptyset$. We show that there exists a number $a \in \mathfrak{m}$ with (487). Because suppose it would be

(488) $$\chi(a) = 1 \qquad (a \in \mathfrak{m})\ .$$

Then let a_1, a_2 two numbers with

(489) $$\langle a_1,q\rangle = \langle a_2,q\rangle = 1$$

and

(490) $$a_1 \equiv a_2 \bmod r\ .$$

Because of (489) there is a c with

(491) $$a_1c \equiv a_2 \bmod q ,$$

hence

(492) $$a_1c \equiv a_2 \bmod r .$$

From (489) till (492) it follows

$$\langle c,q\rangle = 1, \ c \equiv 1 \bmod r ,$$

i. e., $c \in \mathfrak{m}$. From (488), (491) we get

$$\chi(a_2) = \chi(a_1)\chi(c) = \chi(a_1) .$$

Hence from (489), (490) it would follow

(493) $$\chi(a_2) = \chi(a_1) .$$

But then χ would be induced by a character mod r, i. e., χ would not be primitive. But since χ is primitive there must be a number $a \in \mathfrak{m}$ with (487).

With d also ad runs over all residue-classes mod q. Hence to each d there belongs a decomposition

(494) $$ad = k^*r + n^* .$$

Now it follows from (482), (486), (494)

$$n \equiv d \equiv ad \equiv n^* \bmod r ,$$

hence

(495) $$n \equiv n^* \bmod r .$$

Now let k run over all residueclasses mod t. Then

(496) $$a(kr + n) = k^* r + n^*$$

runs over t^m different residueclasses mod q. For these (495) must hold. Hence k^* runs exactly over all residueclasses mod t. Let

(497) $$A = [1,\dots,1,a] .$$

Then it follows

$$\chi(a)G(m,\chi,C) = G(m,\chi,AC) =$$

$$q^{-\frac{m^2}{2}} \sum_{\substack{D_1 \bmod q \\ n \bmod r \\ k \bmod t}} \chi(\mathrm{Det}\begin{pmatrix} D_1 \\ a(kr+n) \end{pmatrix}) \exp(2\pi i\, \frac{c_1 d_1 + \dots + c_{m-1} d_{m-1}}{q} + \frac{bn_m}{r}) =$$

$$q^{-\frac{m^2}{2}} \sum_{\substack{D_1 \bmod q \\ n \bmod r \\ k^* \bmod t}} \chi(\mathrm{Det}\begin{pmatrix} D_1 \\ k^*r+n \end{pmatrix}) \exp(2\pi i\, \frac{c_1 d_1 + \dots + c_{m-1} d_{m-1}}{q} + \frac{bn_m}{r}) =$$

$$G(m,\chi,C) .$$

Hence

(498) $$\chi(a)G(m,\chi,C) = G(m,\chi C) .$$

From (487), (498) we deduce (474) .

Formula (471) was proved in Gričenko [12], page 607 but it also follows from the thetatransformation formula that will be proved later (Theorem 45). Theorem 42 is proved.

DEFINITION 3: Let

(499) $$\chi = (\chi_1, \ldots, \chi_{w^*})$$

be a row of even characters mod q and put

(500) $$\tau_1(1^*, \chi) = \prod_{\mu=1}^{w^*} G(1_\mu, \chi_\mu) ,$$

(501) $$\tau_2(1^*, \chi) = \prod_{\mu=1}^{w^*} G(1, \chi_\mu)^{1_\mu} .$$

DEFINITION 4: Let $\chi = (\chi_1, \ldots, \chi_{w^*})$ be a row of even characters mod q and

(502) $$Y \in \mathcal{Y}(n),\ T \in \mathcal{Y}(n^*) .$$

Set

(503) $$\check{\theta}_\alpha(1, \chi, Y, T) = (\operatorname{Det} Y)^{\frac{n^*}{4}} (\operatorname{Det} T)^{\frac{n}{4}} \times$$

$$\sum_{A = (\overset{1}{A}_{\nu\mu}) \in \mathcal{A}_\alpha(1; w, w^*)} \chi_1(\operatorname{Det} \overset{1}{A}_1) \ldots \chi_{w^*}(\operatorname{Det} \overset{1}{A}_{w^*}) \exp(-\tfrac{\pi}{q} \operatorname{Tr}(Y[AW(n^*)]T))$$

$$(\alpha = 1,2) .$$

Set

(504) $$\overleftarrow{1} = (1, \ldots, 1, 1_w)$$

with n^* times 1 and

(505) $$\overleftarrow{\chi} = (\overleftarrow{\chi}_1, \ldots, \overleftarrow{\chi}_{n^*})$$

with

(506) $$\overleftarrow{\chi}_\iota = \chi_{w(\iota)} \qquad (\iota = 1, \ldots, n^*) .$$

Then

$$(507)\qquad \check{\theta}_2(l,\chi,Y,T) = \check{\theta}_1(\bar{l},\bar{\chi},Y,T)\ .$$

THEOREM 43: Let $q > 1$. Then

$$(508)\qquad \check{\theta}_\alpha(l,\chi,Y,T) = (\operatorname{Det} Y)^{\frac{n^*}{4}}(\operatorname{Det} T)^{\frac{n}{4}} \times \sum_{A = (\overset{l}{A}_{\nu\mu}) \in \mathcal{B}_\alpha(l;w,w^*)} \chi_1(\operatorname{Det} \overset{l}{A}_1) \ldots \chi_{w^*}(\operatorname{Det} \overset{l}{A}_{w^*}) \exp(- \tfrac{\pi}{q} \operatorname{Tr}(Y[AW(n^*)]T))$$

$$(\alpha = 1,2)\ .$$

PROOF: Apply $\chi(a) = 0$ for $\langle a,q\rangle > 1$.

THEOREM 44: Set

$$(509)\qquad \tilde{\chi}_\mu = \chi_{w-\mu}^{-1} \qquad (\mu = 1,\ldots,w^*)$$

and for $U \in \Psi_\alpha(l^*)$ define $\tilde{U}$ by (310) with l^* instead of l .

Then

$$(510)\qquad \check{\theta}_\alpha(l,\chi,Y,T[\tilde{U}]) =$$

$$\Big(\prod_{\nu=1}^{w^*} \chi_\nu(\operatorname{Det} \overset{l^*}{U}_\nu)\Big)\check{\theta}_\alpha(l,\chi,Y,T) = \Big(\prod_{\nu=1}^{w^*} \tilde{\chi}_\nu(\operatorname{Det} \overset{\tilde{l}^*}{\tilde{U}}_\nu)\Big)\check{\theta}_\alpha(l,\chi,Y,T)\ (\alpha=1,2)$$

PROOF: From (310),(503) one gets

$$\check{\theta}_\alpha(l,\chi,Y,T[\tilde{U}]) = (\operatorname{Det} Y)^{\frac{n^*}{4}}(\operatorname{Det} T)^{\frac{n}{4}} \times \sum_{A = (\overset{l}{A}_{\nu\mu}) \in \mathcal{O}_\alpha(l;w,w^*)} \chi_1(\operatorname{Det} \overset{l}{A}_1)\ldots \chi_{w^*}(\operatorname{Det} \overset{l}{A}_{w^*}) \exp(- \tfrac{\pi}{q} \operatorname{Tr}(Y[AU^{-1}W(n^*)]T)).$$

Substituting AU instead of A we get the first of the formulas (510). Using (311) and (509) we get the second. Theorem 44 is proved.

Now we prove the thetafransformation formula from wich (471) follows.

THEOREM 45: Let $\chi_1,\ldots,\chi_{w^*}$ be primitive even characters mod q. Then

$$\check{\theta}_\alpha(1,\chi,Y,T) = \tau_\alpha(1^*,\chi)\check{\theta}_\alpha(\check{1},\check{\tilde{\chi}},\check{\tilde{Y}},\tilde{T}) \qquad (\alpha = 1,2)\,. \tag{511}$$

PROOF: Because of (507) it suffices to consider the case $\alpha = 1$.

Starting with (503) we set

$$A = B + qCW(n^*) \tag{512}$$

with

$$B = (\overset{1}{B}_{\nu\mu}) \in \mathfrak{H}_1(1;w,w^*) \tag{513}$$

and integral C. From (503) follows

$$\begin{aligned}\check{\theta}_1(1,\chi,Y,T) = {} & (\operatorname{Det} Y)^{\frac{n^*}{4}}(\operatorname{Det} T)^{\frac{n}{4}} \times \\ & \sum_{B \bmod q} \chi_1(\operatorname{Det} \overset{1}{B}_1) \ldots \chi_{w^*}(\operatorname{Det} \overset{1}{B}_{w^*}) \times \\ & \sum_{C} \exp(-\pi \operatorname{Tr}((qY)[C + \frac{BW(n^*)}{q}]T))\,.\end{aligned} \tag{514}$$

To the inner sum we apply (457) with qY instead of Y, $V = \frac{BW(n^*)}{q}$, $U = 0$, $m = n^*$.

Then

(515) $$\check{\theta}_1(1,\chi,Y,T) = (\mathrm{Det}\ Y)^{-\frac{n^*}{4}}(\mathrm{Det}\ T)^{-\frac{n}{4}} q^{-\frac{nn^*}{2}} \times$$

$$\sum_C \Big(\sum_{B \bmod q} \chi_1(\mathrm{Det}\ \overset{1}{B}_1) \ldots \chi_{w^*}(\mathrm{Det}\ \overset{1}{B}_{w^*}) \exp(2\pi i\ \mathrm{Tr}\ \frac{C'BW(n^*)}{q})\Big) \times$$

$$\exp(-\frac{\pi}{q}\ \mathrm{Tr}(Y^{-1}[C]T^{-1})) .$$

Here the summation runs over all integral C.

Set

(516) $$A = Q(1)C$$

with $A = (\overset{1}{A}_{\nu\mu})$ and

(517) $$\overset{\check{1}}{A}_{w\mu} \equiv 0 \bmod q \qquad (\mu = 1,\ldots,w^*) .$$

Then from (386), (387), (389), (390), (515), (516) we get

(518) $$\check{\theta}_1(1,\chi,Y,T) = (\mathrm{Det}\ \check{Y})^{\frac{n^*}{4}} (\mathrm{Det}\ \tilde{T})^{\frac{n}{4}} \times$$

$$\sum_A \Phi(A) \exp(-\frac{\pi}{q}\ \mathrm{Tr}(\check{Y}[AW(n^*)]\tilde{T}))$$

with

(519) $$\Phi(A) = q^{-\frac{n^{*2}}{2}} \times$$

$$\sum_{B \bmod q} \chi_1(\mathrm{Det}\ \overset{1}{B}_1) \ldots \chi_{w^*}(\mathrm{Det}\ \overset{1}{B}_{w^*}) \exp(2\pi i\ \mathrm{Tr}(\frac{A'Q(1)^{-1}BW(n^*)}{q}))$$

Let

(520) $$Q(1)^{-1}BW(n^*) = D = (\overset{\check{1}}{D}_{\nu\mu}) .$$

Then

$$(521)\qquad \overset{\vee}{\overset{1}{D}}_{\nu\mu} = 0 \qquad (\nu = w;\ \mu = 1,\dots,w^* \text{ and } 1 \le \nu < \mu \le w^*),$$

$$(522)\qquad \mathrm{Det}\ \overset{1}{B}_{\mu} = \mathrm{Det}\ \overset{\vee}{\overset{1}{D}}_{w-\mu} \qquad (\mu = 1,\dots,w^*)\ .$$

Hence

$$(523)\qquad \mathrm{Tr}\Big(\frac{A'Q(1)^{-1}BW(n^*)}{q}\Big) = \sum_{\mu=1}^{w^*} \mathrm{Tr}\Big(\frac{\overset{\vee}{\overset{1}{A'}}_{\mu}\ \overset{\vee}{\overset{1}{D}}_{\mu}}{q}\Big) + \sum_{1\le\mu<\nu\le w^*} \mathrm{Tr}\Big(\frac{\overset{\vee}{\overset{1}{A'}}_{\nu\mu}\ \overset{\vee}{\overset{1}{D}}_{\nu\mu}}{q}\Big)\ .$$

From (509), (519), (522), (523) we deduce

$$(524)\qquad \Phi(A) = q^{-\frac{n^*}{2}} \prod_{\mu=1}^{w^*}\Big(\sum_{D_\mu \bmod q} \overset{\approx}{\chi}_{\mu}(\mathrm{Det}\ D_\mu)\exp\Big(2\pi i\ \mathrm{Tr}\Big(\frac{A'_\mu D_\mu}{q}\Big)\Big)\Big) \times$$
$$\prod_{1\le\mu<\nu\le w^*}\ \sum_{D_{\nu\mu} \bmod q} \exp\Big(2\pi i\ \mathrm{Tr}\ \frac{A'_{\nu\mu}\ D_{\nu\mu}}{q}\Big)\ .$$

The sum over $D_{\nu\mu}$ is 0 for $A_{\nu\mu} \not\equiv 0 \bmod q$ and it is $q^{\check{1}_\nu \check{1}_\mu}$ for $A_{\nu\mu} \equiv 0 \bmod q$. Hence because of (468), (517) it follows

$$(525)\qquad \Phi(A) = \left\{ \begin{array}{ll} \prod_{\mu=1}^{w^*} G(\check{1}_{\mu}, \overset{\approx}{\chi}_{\mu}, A_{\mu}) & (A \in \mathfrak{A}_1(\check{1};w,w^*))\ , \\ 0 & (A \notin \mathfrak{A}_1(\check{1};w,w^*)) \end{array} \right\} ,$$

and because of theorem 42 and definition 3

$$(526)\qquad \Phi(A) = \tilde{\chi}_1(\mathrm{Det}\ \overset{\vee}{\overset{1}{A}}_1) \dots \tilde{\chi}_{w^*}(\mathrm{Det}\ \overset{\vee}{\overset{1}{A}}_{w^*}) \times$$
$$\left\{ \begin{array}{ll} \tau_1(\widetilde{1^*}, \overset{\approx}{\chi}) & (A \in \mathfrak{A}_1(\check{1};w,w^*)) \\ 0 & (A \notin \mathfrak{A}_1(\check{1};w,w^*)) \end{array} \right\} .$$

From (518), (526) we get

(527) $$\check{\theta}_1(1,\chi,Y,T) = (\mathrm{Det}\ \check{Y})^{\frac{n^*}{4}} (\mathrm{Det}\ \tilde{T})^{\frac{n}{4}} \tau_1(\widetilde{1^*},\tilde{\bar{\chi}}) \times$$
$$\sum_{A \in \mathfrak{A}_1(\check{1};w,w^*)} \tilde{\chi}_1(\mathrm{Det}\ \check{A}_1) \ldots \tilde{\chi}_{w^*}(\mathrm{Det}\ \check{A}_{w^*}) \times$$
$$\exp(-\frac{\pi}{q}\,\mathrm{Tr}(\check{Y}[AW(n^*)]\tilde{T}) = \tau_1(\widetilde{1^*},\tilde{\bar{\chi}})\check{\theta}_1(\check{1},\tilde{\chi},\check{Y},\tilde{T}) .$$

But

(528) $$\tau_1(\widetilde{1^*},\tilde{\bar{\chi}}) = \tau_1(1^*,\chi) .$$

From (527), (528) we get (511). Theorem 45 is proved.

Let $D^*(k,T)$ be defined by (403) and set

(529) $$D(q,T) = \left\{ \begin{array}{ll} D^*(n,T) & (q = 1) \\ & \\ 1 & (q > 1) \end{array} \right\} .$$

This is a generalization of (22). From (405) we get

(530) $$D(q,T) = D(q,T^{-1}) = D(q,\tilde{T}) .$$

Set

(531) $$\theta_\alpha(1,\chi,Y,T) = D(q,T)\check{\theta}_\alpha(1,\chi,Y,T) \qquad (\alpha = 1,2) .$$

THEOREM 46: Let $U \in \Psi_\alpha(1^*)$. Then

(532) $$\theta_\alpha(1,\chi,Y,T[\tilde{U}]) =$$
$$(\prod_{\nu=1}^{w^*} \chi_\nu(\mathrm{Det}\ \overset{1^*}{U_\nu}))\theta_\alpha(1,\chi,Y,T) = (\prod_{\nu=1}^{w^*} \tilde{\chi}_\nu(\mathrm{Det}\ \overset{\widetilde{1^*}}{\tilde{U}_\nu}))\theta_\alpha(1,\chi,Y,T) \quad (\alpha = 1,2).$$

PROOF: Apply theorem 44 and the invariance of $D(q,T)$.

THEOREM 47: Let $\chi_1,\ldots,\chi_{w*}$ be primitive even characters mod q. Then

$$\theta_\alpha(1,\chi,Y,T) = \tau_\alpha(1^*,\chi)\theta_\alpha(\check{1},\tilde{\chi},\check{Y},\tilde{T}) \qquad (\alpha = 1,2) . \tag{533}$$

PROOF: Apply theorem 45 and (530).

THEOREM 48: Let C be a $n^* \times n^*$ matrix. Then

$$(\mathrm{Det}\ \tfrac{\partial}{\partial T})\exp(\mathrm{Tr}(CT)) = (\mathrm{Det}\ C)\exp(\mathrm{Tr}(CT)) . \tag{534}$$

PROOF: Easy computation.

THEOREM 49: Let $q = 1$. Then

$$\theta_2(1,\chi,Y,T) = \theta_1(1,\chi,Y,T) = (\mathrm{Det}\ Y)^{\frac{n^*}{4}}(\mathrm{Det}\ T)^{\frac{n}{4}} \times \tag{535}$$

$$\sum_{A \in \mathcal{L}_1(1;w,w^*)} P_{\frac{n}{2}}(T)\exp(-\pi\ \mathrm{Tr}(Y[AW(n^*)]T)) .$$

PROOF: For $q = 1$, $\mathfrak{A}_1(1;w,w^*) = \mathfrak{A}_2(1;w,w^*)$. Hence the first equality follows. From (403), (503) we see that $\theta_1(1,\chi,Y,T)$ is the sum on the right-hand-side of (535) taken over $\mathfrak{A}_1(1;w,w^*)$. From (399), (400) we see that $P_{\frac{n}{2}}(T) = \ldots (\mathrm{Det}\ \frac{\partial}{\partial T})$. Hence because of (534) all summands with $\mathrm{Rk}\ A < n^*$ are zero. So it suffices to take the summation over the A with $\mathrm{Rk}\ A = n^*$. Theorem 49 is proved.

In [33], page 79 Maaß considers the operator $L = (\mathrm{Det}\ Y)^h(\mathrm{Det}\ \frac{\partial}{\partial T})^h$ and then computes $\hat{L}$. Applying this with $h = 1$ we get

$$\hat{M}_{n*}(T) = (-1)^{n^*}(\mathrm{Det}\ T)^{\frac{n^*+1}{2}}(\mathrm{Det}\ \tfrac{\partial}{\partial T})(\mathrm{Det}\ T)^{\frac{1-n^*}{2}} . \tag{536}$$

Combining (399), (400), (536) we obtain

$$P_k(T) = (-1)^{n^*}(\mathrm{Det}\ T)^{\frac{n^*+1}{2}-k}(\mathrm{Det}\ \tfrac{\partial}{\partial T})(\mathrm{Det}\ T)^{\frac{3-n^*}{2}+k}(\mathrm{Det}\ \tfrac{\partial}{\partial T}) . \tag{537}$$

THEOREM 50: Let

$$X \in \mathcal{Y}(n) . \tag{538}$$

Then

$$(\mathrm{Det}\,\tfrac{\partial}{\partial X})(f(l,X^{-1},s)\exp(-\mathrm{Tr}\,X)) = f(l,X^{-1},s+e(w))\exp(-\mathrm{Tr}X)R(X), \tag{539}$$

where $R(X)$ is a polynomial of degree $\leq n$ and $e(w) = (1,\ldots,1)$ with w times 1.

PROOF: Because of (432) it suffices to prove the theorem for $l = \overset{\circ}{l}$. Suppose at first

$$\sigma_\nu > \frac{n-1}{4} \qquad (\nu = 1,\ldots,n), \tag{540}$$

and set

$$\gamma(s) = \pi^{\frac{n(n-1)}{4}} \prod_{\nu=1}^{n} \Gamma(s_\nu - \tfrac{n-1}{4}) . \tag{541}$$

Then because of (437)

$$f(\overset{\circ}{l},X^{-1},s) = \gamma^{-1}(s)\mathcal{J}(\overset{\circ}{l},X^{-1},s) , \tag{542}$$

because of (435)

$$f(\overset{\circ}{l},X^{-1},s)\exp(-\mathrm{Tr}\,X) = \gamma^{-1}(s) \int_{Y \in \mathcal{Y}(n)} f(l,Y,s)\exp(-\mathrm{Tr}(X(Y+E)))dv_Y. \tag{543}$$

Hence because of (534)

$$(\mathrm{Det}\,\tfrac{\partial}{\partial X})(f(\overset{\circ}{l},X^{-1},s)\exp(-\mathrm{Tr}\,X)) = \tag{544}$$

$$(-1)^n\gamma^{-1}(s) \int_{Y \in \mathcal{Y}(n)} f(\overset{\circ}{l},Y,s)\mathrm{Det}(Y+E)\exp(-\mathrm{Tr}(X(Y+E)))dv_Y =$$

$$(-1)^n\gamma^{-1}(s)\exp(-\mathrm{Tr}\,X) \int_{Y \in \mathcal{Y}(n)} f(\overset{\circ}{l},Y,s)\mathrm{Det}(Y+E)\exp(-\mathrm{Tr}(XY))dv_Y .$$

With T of type (439) put $X = T^{-1}T'^{-1}$ and make the substitution $Y \to Y[T]$. Then

$$(\mathrm{Det}\ \tfrac{\partial}{\partial X})(f(\overset{\circ}{1},X^{-1},s)\exp(-\mathrm{Tr}\ X)) = (-1)^n \gamma^{-1}(s)\exp(-\mathrm{Tr}\ X) \times$$

$$\int_{Y \in \mathcal{Y}(n)} f(\overset{\circ}{1},Y[T],s)\mathrm{Det}(Y[T] + E)\exp(-\mathrm{Tr}\ Y)dv_Y .$$

Apply (426), (444). Then

$$(545)\quad (\mathrm{Det}\ \tfrac{\partial}{\partial X})(f(\overset{\circ}{1},X^{-1},s)\exp(-\mathrm{Tr}\ X)) = f(\overset{\circ}{1},X^{-1},s+e(n))\exp(-\mathrm{Tr}\ X)\,R^*(s,X)$$

with

$$(546)\quad R^*(s,X) = (-1)^n\gamma^{-1}(s) \int_{Y \in \mathcal{Y}(n)} f(\overset{\circ}{1},Y,s)\mathrm{Det}(Y+X)\exp(-\mathrm{Tr}\ Y)dv_Y .$$

Obviously $R^*(s,X)$ is a polynomial in X of degree $\leq n$.

The function

$$(547)\qquad R(X) = R(s,X) =$$

$$f^{-1}(\overset{\circ}{1},X^{-1},s + e(n))\exp(\mathrm{Tr}\ X)(\mathrm{Det}\ \tfrac{\partial}{\partial X})(f(\overset{\circ}{1},X^{-1},s)\exp(-\mathrm{Tr}\ X))$$

is holomorphic in s for all $s \in \mathbb{C}^n$.

For (540) we have

$$(548)\qquad R(s,X) = R^*(s,X) .$$

Let Φ be a partial derivative of X of order $> n$. Then for (540) we get from (548)

$$(549)\qquad \Phi R(s,X) = 0$$

identically in s and X. By the principle of analytic continuation this must be true for all $s \in \mathbb{C}^n$. Hence $R(s,X)$ is a polynomial in X of degree $\leq n$. Theorem 50 is proved.

Let $a \in \mathbb{C}$. If we apply theorem 50 for the special case $w = 1$ and $T \in \mathcal{Y}(n^*)$ instead of $X \in \mathcal{Y}(n)$ we get for $a \in \mathbb{C}$

(550) $(\mathrm{Det}\ \frac{\partial}{\partial T})((\mathrm{Det}\ T)^a \exp(-\mathrm{Tr}\ T)) = (\mathrm{Det}\ T^{a-1} \exp(-\mathrm{Tr}\ T)R(T),$

where $R(T)$ is a polynomial of degree $\leq n^*$.

From (534), (537), (550) we deduce

(551) $$P_k(T)\exp(-\mathrm{Tr}\ T) = (\mathrm{Det}\ T)R(T)\exp(-\mathrm{Tr}\ T).$$

Here $R(T)$ is a polynomial of degree $\leq n^*$.

THEOREM 51: Let

(552) $$S \in \mathcal{Y}(n^*) .$$

Then there exists a constant $c_{12} = c_{12}(k,n^*) \geq 1$ with

(553) $$P_k(T)\exp(-\mathrm{Tr}(ST)) \leq c_{12}\exp(-\tfrac{1}{2}\mathrm{Tr}(ST)).$$

PROOF: Because of (551) there is a constant c_{12} with

(554) $$P_k(T)\exp(-\mathrm{Tr}\ T) \leq c_{12} \exp(-\tfrac{1}{2}\mathrm{Tr}\ T) .$$

Let A be a real non-singular $n^* \times n^*$ matrix. Since $P_k(T)$ is an invariant operator (554) gives us

$$P_k(T) \exp(-\mathrm{Tr}(T[A])) \leq c_{12}\exp(-\tfrac{1}{2}\mathrm{Tr}(T[A])) .$$

Hence

(555) $$P_k(T) \exp(-\mathrm{Tr}(AA'T)) \leq c_{12} \exp(-\tfrac{1}{2}\mathrm{Tr}(AA'T)) .$$

Since S may be written as $S = AA'$ with suitable A the assertition (553) follows from (555). Theorem 51 is proved.

THEOREM 52: Let

(556) $$q^* = \begin{cases} 2 & (q = 1) \\ q & (q > 1) \end{cases} .$$

There exists a constant $c_{13} = c_{13}(l) \geq 1$ with

(557) $$\operatorname{abs} \theta_\alpha(l,\chi,Y,T) \leq c_{13}(\operatorname{Det} Y)^{\frac{n^*}{4}}(\operatorname{Det} T)^{\frac{n}{4}} \times \sum_{\operatorname{Rk} A = n^*} \exp(-\tfrac{\pi}{q^*} \operatorname{Tr}(Y[A]T) \qquad (\alpha = 1,2) .$$

The summation is over all integral $n \times n^*$ matrices A with $\operatorname{Rk} A = n^*$.

PROOF: Apply theorems 25, 43, 51, formula (535) and write A instead of $AW(n^*)$.

THEOREM 53: Let R_n, R_{n^*} non-singular rational $n \times n$ respectively $n^* \times n^*$ matrices, $\chi_1,\dots,\chi_{w^*}$ arbitrary even characters mod q, $Y \in \mathcal{Y}(n)$ and $j(Y)$ a positive number with

(558) $$Y \geq j(Y)E ,$$

furthermore

(559) $$T \in \mathcal{T}(n^*,\mu) .$$

Then there exists a real number

(560) $$c_{14} = c_{14}(n,n^*,\mu,R_n,R_{n^*}) \geq 1$$

with

$$(561)\qquad \operatorname{abs} \theta_\alpha(1,\chi,Y[R_n],T[R_{n*}]) \le$$
$$c_{14} j(Y)^{-\frac{nn^*}{2}} (\operatorname{Det} Y)^{\frac{n^*}{4}} (\operatorname{Det} T)^{-\frac{n}{4}} \exp(-c_{14}^{-1}\, j(Y) \operatorname{Tr} T) \quad (\alpha = 1,2).$$

PROOF: From (366), (559) we deduce

$$(562)\qquad c_{10}^{-1} \operatorname{Dg} T \le T \le c_{10} \operatorname{Dg} T .$$

Set $R_n = g_n^{-1} G_n$, $R_{n*} = g_{n*}^{-1} G_{n*}$ with $g_n, g_{n*} \in \mathbb{N}$ and integral G_n, G_{n*}. Then from theorem 52 and the inequalities (558), (562) we obtain

$$(563)\qquad \operatorname{abs} \theta_\alpha(1,\chi,Y[R_n],T[R_{n*}]) \le c_{13} (\operatorname{Det} Y)^{\frac{n^*}{4}} (\operatorname{Det} T)^{\frac{n}{4}} \times$$
$$\sum_{\operatorname{Rk} A = n^*} \exp(-2\pi \operatorname{Tr}(\frac{j(Y)}{2q^* g_n^2 g_{n*}^2 c_{10}} (\operatorname{Dg} T)[G_{n*} A' G_n'])) .$$

Form the diagonal matrix

$$(564)\qquad S = \frac{j(Y) \operatorname{Dg} T}{2q^* g_n^2 g_{n*}^2 c_{10}} = [s_1,\dots,s_{n*}]$$

and put

$$(565)\qquad \Phi(n,n^*;s) = \sum_{\operatorname{Rk} B = n^*} \exp(-2\pi \operatorname{Tr}(S[B])) ,$$

where B runs over all integral $n^* \times n$ matrices of $\operatorname{Rk} B = n^*$. Then from (563) we get

$$(566)\quad \operatorname{abs} \theta_\alpha(1,\chi,Y[R_n],T[R_{n*}]) \le c_{13} (\operatorname{Det} Y)^{\frac{n^*}{4}} (\operatorname{Det} T)^{\frac{n}{4}} \Phi(n,n^*;S).$$

If we can prove

$$(567)\qquad \Phi(n,n^*;S) \le d_1 (\operatorname{Det} S)^{-\frac{n}{2}} \exp(-d_1^{-1} \operatorname{Tr} S)$$

with some constant $d_1 \ge 1$, the assertion (561) follows from (562), (564), (566), (567).

Since S is a diagonal matrix

(568) $$\Phi(n,n^*;S) \leq \prod_{\mu=1}^{n^*} \Phi(n,1,s_\mu) .$$

Let $u > 0$ and $\vartheta(u)$ be defined by (31). Then

(569) $$\Phi(n,1,u) \leq n(\vartheta(2u)-1)\vartheta^{n-1}(2u).$$

Applying

(570) $$\vartheta(2u) - 1 \leq \exp(-\pi u)\vartheta(u)$$

and

(571) $$\vartheta(2u) \leq \vartheta(u)$$

we get

(572) $$\Phi(n,1,u) \leq n \exp(-\pi u)\vartheta^n(u) .$$

Let $\varepsilon > 0$ be given. From (35), (572) we get

(573) $$\Phi(n,1,u) \leq d_3 u^{-\frac{n}{2}} \exp(-(\pi - n\varepsilon)u)$$

with some constant $d_3 = d_3(\varepsilon) \geq 1$. Now (567) follows from (568), (573). Theorem 53 is proved.

THEOREM 54: Let $\chi_1,\dots,\chi_{w^*}$ be primitive even characters mod q and set

(574) $$\hat{\chi}_1 = \chi_{w^*}, \ \hat{\chi}_\nu = \chi_{\nu-1} \quad (\nu = 1,\dots,w^*-1) ;$$

(575) $$\hat{\chi} = (\hat{\chi}_1,\dots,\hat{\chi}_{w^*}) .$$

Then

(576) $$\theta_1(1,\chi,Y,T[P^{-1}(1^*)]) =$$

$$q^{-\frac{k_{w-2}\, l_w}{2}} \sum_{\rho=1}^{r} \theta_1(\hat{l},\hat{\chi},Y[K_\rho(l)])\ ,T)\ ,$$

(577) $$\theta_1(\check{l},\tilde{\chi},\check{Y},T[P(l^*)]) = q^{-\frac{k_{w-2}\, l_w}{2}} \sum_{\rho=1}^{r} \theta_1(\check{\hat{l}},\tilde{\hat{\chi}},Y[K_\rho(l)])^{\vee}T)\ .$$

PROOF: Let $q = 1$. Then $r = 1$, $K_1(l) \in \Omega(n)$, $P(l^*) \in \Omega(n^*)$. Then (576) is true because from (535) we deduce that $\theta_1(l,\chi,Y,T)$ is invariant under $Y \to Y[U]$, $T \to T[V]$ with $U \in \Omega(n)$, $V \in \Omega(n^*)$.

Now let $q > 1$. From (330), (332), (340), (344), (345), (391), (508), (529), (531), (574) follows

(578) $$\theta_1(l,\chi,Y,T[P^{-1}(l^*)]) = (\mathrm{Det}\ Y)^{\frac{n^*}{4}} (\mathrm{Det}\ T)^{\frac{n}{4}} \times$$

$$\sum_{A=(\overset{l}{A}_{\nu\mu}) \in \mathcal{L}_1(l;w,w^*)} \chi_1(\mathrm{Det}\ \overset{l}{A}_1)\ldots\chi_{w^*}(\mathrm{Det}\ \overset{l}{A}_{w^*})\exp(-\tfrac{\pi}{q}\,\mathrm{Tr}(Y[AW(n^*)P'^{-1}(l^*)W(n^*)W(n^*)]T))$$

$$= (\mathrm{Det}\ Y)^{\frac{n^*}{4}} (\mathrm{Det}\ T)^{\frac{n}{4}} \times$$

$$\sum_{A=(\overset{l}{A}_{\nu\mu}) \in \mathcal{L}_1(l;w,w^*)} \chi_1(\mathrm{Det}\ \overset{l}{A}_1)\ldots\ \chi_{w^*}(\mathrm{Det}\ \overset{l}{A}_{w^*})\exp(-\tfrac{\pi}{q}\,\mathrm{Tr}(Y[AP^{-1}(l^*)W(n^*)]T))$$

$$= q^{-\frac{k_{w-2}l_w}{2}} \sum_{\rho=1}^{r} (\mathrm{Det}\ Y[K_\rho])^{\frac{n^*}{4}} (\mathrm{Det}\ T)^{\frac{n}{4}} \times$$

$$\sum_{A^*=(\overset{\hat{l}_*}{A^*_{\nu\mu}}) \in \mathcal{L}_1(\hat{l},w,w^*)} \hat{\chi}_1(\mathrm{Det}\ \overset{\hat{l}_*}{A^*_1})\ldots\ \hat{\chi}_{w^*}(\mathrm{Det}\ \overset{\hat{l}_*}{A^*_{w^*}})\exp(-\tfrac{\pi}{q}\,\mathrm{Tr}(Y[K_\rho A^* W(n^*)]T))$$

$$= q^{-\frac{k_{w-2}\, l_w}{2}} \sum_{\rho=1}^{r} \theta_1(\hat{l},\hat{\chi},Y[K_\rho],T)\ .$$

This proves (576).

From (244), (533), (576) we deduce

$$(579) \qquad \tau_1(l^*,\chi)\theta_1(\check{l},\tilde{\chi},\check{Y},\widetilde{T[P^{-1}(l^*)]}) = \tau_1(\mathfrak{z}^* l^*,\hat{\chi}) \times$$

$$q^{-\frac{k_{w-2}\, l_w}{2}} \sum_{\rho=1}^{r} \theta_1(\check{\hat{l}},\tilde{\hat{\chi}},(Y[K_\rho(l)])^\vee, \tilde{\check{T}}) .$$

From (242), (471), (500), (574), (575) we get

$$(580) \qquad \tau_1(\mathfrak{z}^* l^*,\hat{\chi}) = \tau_1(l^*,\chi) \neq 0.$$

Formula (332) gives

$$(581) \qquad \widetilde{T[P^{-1}(l^*)]} = \tilde{T}[P(l^*)] .$$

Put (581) in the left-hand-side of (579), write T instead of $\tilde{T}$ and use (580). Then (577) follows. Theorem 54 is proved.

CHAPTER III. SELBERG'S ZETA- AND L-SERIES

The Selberg's zetafunction in this chapter are identical with the zetafunctions considered in Maaβ [33], § 17 and Terras [45], [46]. Selberg's L-series are associated to these zetafunctions in the same way as Dirichlet's L-series are associated with Riemann's zetafunction. We prove analytic continuation and functional equations of these functions. The methods are similar to those of Maaβ [33], § 17.

---

§ 8. DESCENDING CHAINS

A descending chain is a system of matrices with integral elements such that the number of rows and columns become smaller by going down the chain. It is defined when two descending chains are equivalent and equivalence class invariants are defined. The theory of descending chains is developed so far as it is needed for Selberg's zetafunctions and L-series.

DEFINITION 5: Let $\alpha = 1,2$. A matrix system

(582) $$\mathfrak{A} = \{A_{w-1},\dots,A_1\}$$

with

(583) $$A_\beta \in \mathscr{L}_\alpha(1;\beta+1,\beta) \qquad (\beta = 1,\dots,w-1),$$

is called a "descending chain" of type a. Two descedning chains $\mathfrak{A}$ and

(584) $$\mathfrak{A}^* = \{A^*_{w-1},\dots,A^*_1\}$$

are called "equivalent", if there exist

(585) $$U_\beta \in \Psi_\alpha((l_1,\dots,l_\beta)) \qquad (\beta = 1,\dots,w-1)$$

with

(586) $$A^*_{w-1} = A_{w-1}\, U_{w-1} \;,$$

(587) $$A^*_\beta = U^{-1}_{\beta+1}\, A_\beta\, U_\beta \qquad (\beta = 1,\dots,w-2) \;.$$

The equivalence class to which $\mathfrak{A}$ belongs is $\{\mathfrak{A}\}$.

THEOREM 55: Set

(588) $$A_\beta = (\overset{1}{A}_{\beta,\nu\mu}),\; A^*_\beta = (\overset{1*}{A}_{\beta,\nu\mu}) \qquad (\beta = 1,\dots,w-1),$$

(589) $$B_{\beta\gamma} = A_\beta\, A_{\beta-1} \cdots A_\gamma = (\overset{1}{B}_{\beta\gamma,\nu\mu}) \in \mathscr{B}_\alpha(1;\beta+1,\gamma) \qquad (1 \le \gamma \le \beta \le w-1) \;,$$

(590) $$B^*_{\beta\gamma} = A^*_\beta\, A^*_{\beta-1} \cdots A^*_\gamma = (\overset{1*}{B}_{\beta\gamma,\nu\mu}) \in \mathscr{B}_\alpha(1;\beta+1,\gamma) \qquad (1 \le \gamma \le \beta \le w-1) \;.$$

Then

(591) $$B^*_{w-1,\gamma} = B_{w-1,\gamma}\, U_\gamma \qquad (\gamma = 1,\dots,w-1) \;,$$

(592) $$B^*_{\beta\gamma} = U^{-1}_{\beta+1}\, B_{\beta\gamma}\, U_\gamma \qquad (1 \le \gamma \le \beta \le w-2) \;.$$

The following natural numbers are invariant under equivalence:

(593) $$\langle B_{\beta\gamma}\rangle \qquad (1 \le \gamma \le \beta \le w-1) \;,$$

(594) $$h_{11} = h_{11}(\mathfrak{A}) = \langle B_{11}\rangle \;,$$

(595) $$h_{\beta 1} = h_{\beta 1}(\mathfrak{A}) = \frac{\langle B_{\beta 1}\rangle}{\langle B_{\beta-1,1}\rangle} \qquad (\beta = 2,\dots,w-1) \;,$$

(596) $$h_{\beta\beta} = h_{\beta\beta}(\mathfrak{A}) = \frac{\langle B_{\beta\beta}\rangle\langle B_{\beta-1,\beta-1}\rangle}{\langle B_{\beta,\beta-1}\rangle} \qquad (\beta = 2,\dots,w-1) \;,$$

(597) $$h_{\beta\gamma} = h_{\beta\gamma}(\mathfrak{A}) = \frac{\langle B_{\beta\gamma}\rangle\langle B_{\beta-1,\gamma-1}\rangle}{\langle B_{\beta-1,\gamma}\rangle\langle B_{\beta,\gamma-1}\rangle} \qquad (2 \le \gamma < \beta \le w-1) \;.$$

Set

$$(598)\qquad B_{w-1,\gamma} = \binom{C_\gamma}{*}$$

with a $k_\gamma \times k_\gamma$ matrix C_γ. Then also

$$(599)\qquad \hat{h}_\gamma = \hat{h}_\gamma(\mathfrak{A}) = \text{abs } C_\gamma \qquad (\gamma = 1,\dots,w-1),$$

$$(600)\qquad \check{h}_\gamma = \check{h}_\gamma(\mathfrak{A}) = \frac{\hat{h}_\gamma}{\langle B_{w-1,\gamma}\rangle} \qquad (\gamma = 1,\dots,w-1)$$

are class invariants lying in $\mathbb{N}$. All those class invariants are coprime to q. Hence by

$$(601)\qquad h_1^* \equiv \check{h}_1 \bmod q;\ \check{h}_\gamma \equiv h_\gamma^* \check{h}_{\gamma-1} \bmod q \qquad (\gamma = 2,\dots,w-1)$$

one can define w-1 class invariants

$$(602)\qquad (h_\gamma^* \bmod q) = (h_\gamma^*(\mathfrak{A}) \bmod q) \qquad (\gamma = 1,\dots,w-1).$$

Furthermore

$$(603)\qquad \langle A_\beta\rangle = \prod_{\nu=1}^{\beta} h_{\beta\nu} \qquad (\beta = 1,\dots,w-1),$$

$$(604)\qquad \langle B_{\beta\gamma}\rangle = \prod_{\mu=1}^{\gamma} \prod_{\nu=\gamma}^{\beta} h_{\nu\mu} \qquad (1 \leq \gamma \leq \beta \leq w-1),$$

$$(605)\qquad \overset{1}{B}_{\beta\gamma,\nu} \equiv \overset{1}{A}_{\beta,\nu} \cdots \overset{1}{A}_{\gamma,\nu} \bmod q \, (1 \leq \nu \leq \gamma \leq \beta \leq w-1),$$

$$(606)\qquad \text{Det } C_\gamma \equiv \prod_{\mu=1}^{\gamma} \prod_{\nu=\gamma}^{w-1} \text{Det } \overset{1}{A}_{\nu\mu} \bmod q \qquad (\gamma = 1,\dots,w-1).$$

PROOF: Formula (605) follows from (589). From (583), (605) one easily gets $B_{\beta\gamma} \in \mathcal{L}_\alpha(1;\beta+1,\gamma)$. Herewith (589) is completely proved. Obviously also (590) is true. The formulas (591), (592) follow from (586), (587), (589), (590). The formulas (591), (592) show that $\langle B_{\beta\gamma}\rangle$ are equivalence-class invariants. Because of

(583) is $\langle B_{\beta\gamma}\rangle \neq 0$, hence $\langle B_{\beta\gamma}\rangle \in \mathbb{N}$. Hence the $h_{\beta\gamma}$ $(1 \leq \gamma \leq \beta \leq w-1)$ are positive equivalence class invariants. Because of theorem 31 they ly in $\mathbb{N}$. From (591), (598) one sees that also (599) is an equivalence class invariant lying in $\mathbb{N}$. $\check{h}_\gamma$ is a positive equivalence class invariant. From the definition of $\langle B_{w-1,\gamma}\rangle$ follows $\langle B_{w-1,\gamma}\rangle | \text{abs } C_\gamma$ hence $\check{h}_\gamma \in \mathbb{N}$.

From the definition of $\mathcal{S}_\alpha(1;\beta+1,\gamma)$ it follows that all those class invariants are coprime to q. From (599) follows

(607) $$\hat{h}_\gamma = \pm \text{ Det } C_\gamma \qquad (\gamma = 1,\ldots,w-1) .$$

From (589), (598) follows

(608) $$\text{Det } C_\gamma \equiv \prod_{\iota=1}^{\gamma} \text{Det } B_{w-1,\gamma,\iota} \text{ mod } q$$

and from this and (605) we get (606). Finally (603) is a special case of (604). Hence it suffices to prove (604).

For $\beta = \gamma = 1$ follows (604) from (594). Now let $\gamma = 1$. Then (604) is identical with

(609) $$\langle B_{\beta 1}\rangle = \prod_{\varkappa=1}^{\beta} h_{\beta 1} .$$

This follows from (595) by induction with respect to β. Now let $1 < \gamma \leq w-1$ and the formula (604) already proved for all $\gamma^* < \gamma$ and β arbitrary in the interval $\gamma^* \leq \beta \leq w-1$. We shall prove (604) for γ and all β with $\gamma \leq \beta \leq w-1$. First let $\beta = \gamma$. From (596) and the induction assumption follows

$$\langle B_{\gamma\gamma}\rangle = h_{\gamma\gamma}\langle B_{\gamma,\gamma-1}\rangle\langle B_{\gamma-1,\gamma-1}\rangle^{-1} =$$

$$h_{\gamma\gamma}\Big(\prod_{\iota=1}^{\gamma-1}\ \prod_{\varkappa=\gamma-1}^{\gamma} h_{\varkappa\iota}\Big)\Big(\prod_{\iota=1}^{\gamma-1} h_{\gamma-1,\iota}^{-1}\Big) =$$

$$h_{\gamma\gamma} \prod_{\iota=1}^{\gamma-1} h_{\gamma\iota} = \prod_{\iota=1}^{\gamma} h_{\gamma\iota} .$$

Herewith one has (604) for $\beta = \gamma$. Now let $\gamma < \beta \leq w-1$ and assume that the assertion is true for all β^* with $\gamma \leq \beta^* < \beta$. From (597) and the induction assumption we get

$$\langle B_{\beta\gamma}\rangle = h_{\beta\gamma}\langle B_{\beta-1,\gamma}\rangle\langle B_{\beta,\gamma-1}\rangle\langle B_{\beta-1,\gamma-1}\rangle^{-1} =$$

$$h_{\beta\gamma}\Big(\prod_{\iota=1}^{\gamma}\prod_{\kappa=\gamma}^{\beta-1} h_{\iota\kappa}\Big)\Big(\prod_{\iota=1}^{\gamma-1}\prod_{\kappa=\gamma-1}^{\beta} h_{\kappa\iota}\Big)\Big(\prod_{\iota=1}^{\gamma-1}\prod_{\kappa=\gamma-1}^{\beta-1} h_{\kappa\iota}^{-1}\Big) =$$

$$\Big(\prod_{\iota=1}^{\gamma}\prod_{\kappa=\gamma}^{\beta} h_{\kappa\iota}\Big)\Big(\prod_{\iota=1}^{\gamma-1} h_{\beta\iota}^{-1}\Big)\Big(\prod_{\iota=1}^{\gamma-1}\prod_{\kappa=\gamma-1}^{\beta} h_{\kappa\iota}\Big)\Big(\prod_{\iota=1}^{\gamma-1}\prod_{\kappa=\gamma-1}^{\beta-1} h_{\kappa\iota}^{-1}\Big) =$$

$$\Big(\prod_{\iota=1}^{\gamma}\prod_{\kappa=\gamma}^{\beta} h_{\kappa\iota}\Big)\Big(\prod_{\iota=1}^{\gamma-1}\prod_{\kappa=\gamma-1}^{\beta} h_{\kappa\iota}\Big)\Big(\prod_{\iota=1}^{\gamma-1}\prod_{\kappa=\gamma-1}^{\beta} h_{\kappa\iota}^{-1}\Big) =$$

$$\prod_{\iota=1}^{\gamma}\prod_{\kappa=\gamma}^{\beta} h_{\kappa\iota} .$$

Herewith (604) and theorem 55 are proved.

DEFINITION 6: A descending chain is called "special" if

(610) $$A_\beta \in \mathcal{J}^*(1;\beta+1,\beta) \qquad (\beta = 1,\ldots,w-1) .$$

The set of special descending chains may be denoted by $\mathcal{G}(1)$. A descending chain is called "reduced" if there exists a

(611) $$U \in \mathcal{U}(1),$$

such that

(612) $$\{U^{-1}A_{w-1}, A_{w-2}, \ldots, A_1\} \in \mathcal{G}(1) .$$

The set of reduced descending chains may be denoted with $\mathcal{R}(1)$.

THEOREM 56: If

(613) $$\mathfrak{A} = \{UD_{w-1}, D_{w-2}, \ldots, D_1\} \in \mathcal{R}(1)$$

with

(614) $$U = (\overset{1}{U}_{\nu\mu}) \in \check{\mathcal{U}}(1),$$

(615) $$D_\beta = (\overset{1}{D}_{\beta,\nu\mu}) \in \mathcal{J}^*(1;\beta+1,\beta) \qquad (\beta = 1,\ldots,w-1),$$

we have

(616) $$h_{\beta\gamma}(\mathfrak{A}) = \operatorname{Det} \overset{1}{D}_{\beta,\gamma} \qquad (1 \leq \gamma \leq \beta \leq w-1),$$

(617) $$\hat{h}_\gamma(\mathfrak{A}) \equiv \pm \prod_{\mu=1}^{\gamma} (\operatorname{Det} \overset{1}{U}_\mu \prod_{\nu=\gamma}^{w-1} \operatorname{Det} \overset{1}{D}_{\nu\mu}) \bmod q \quad (\gamma=1,\ldots,w-1),$$

(618) $$\check{h}_\gamma(\mathfrak{A}) \equiv \pm \prod_{\mu=1}^{\gamma} (\operatorname{Det} \overset{1}{U}_\mu) \bmod q \qquad (\gamma=1,\ldots,w-1),$$

(619) $$h^*_\gamma(\mathfrak{A}) \equiv \pm \operatorname{Det} \overset{1}{U}_\gamma \bmod q \qquad (\gamma=1,\ldots,w-1).$$

If $\mathfrak{A} \in \mathcal{G}(1)$ one may instead of (618), (619) even say

(620) $$\hat{h}_\gamma(\mathfrak{A}) = \prod_{\mu=1}^{\gamma} \prod_{\nu=\gamma}^{w-1} (\operatorname{Det} \overset{1}{D}_{\nu\mu}) \qquad (\gamma=1,\ldots,w-1),$$

(621) $$\check{h}_\gamma(\mathfrak{A}) = 1 \qquad (\gamma=1,\ldots,w-1).$$

PROOF: The formulas (616) follow from theorem 31 and the equations (594) till (597). Formula (617) follows from (599), (606), (614) (615). Formula (618) is a consequence of (600), (615), (617). Formula (619) follows from (601),(618). Formulas (620), (621) follow from (599), (600) and (615). Theorem 56 is proved.

THEOREM 57: Each descending chain of type 2 is also of type 1. Each descending chain of type 1 is equivalent to a descending chain of type 2. Each reduced descending chain is of type 2. Each equivalence class of type α $(\alpha=1,2)$ contains exactly one reduced descending chain. Formaly one has the Cartesian product decomposition

$$(622) \qquad \mathcal{R}(1) = \breve{\mathcal{M}}(1) \times \mathcal{Q}(1).$$

PROOF: Because of

$$(623) \qquad \mathcal{B}_2(1;\nu,\mu) \subset \mathcal{B}_1(1;\nu,\mu) \qquad (1 \leq \mu,\ \nu \leq w)$$

and (582) one sees that each descending chain of type 2 is also of type 1.

From

$$(624) \qquad \breve{\mathcal{M}}(1) \in \Psi_2(1)$$

and

$$(625) \qquad \mathcal{J}^*(1;\nu,\mu) \subset \mathcal{J}_2(1;\nu,\mu) \qquad (1 \leq \mu \leq \nu \leq w)$$

it follows that each reduced descending chain is of type 2. If we can show that each descending chain of type 1 is equivalent to a reduced descending chain, we have also proved that each descending chain of type 1 is equivalent to a descending chain of type 2.

Let $\alpha = 1,2$ and

$$(626) \qquad \mathfrak{A}^{**} = \{A_{w-1}^{**},\dots,A_1^{**}\}$$

be a descending chain of type α. By

$$(627) \qquad A_{w-1}^{*} = A_{w-1}^{**}\ V_{w-1}^{*}\ ,$$

$$(628) \qquad A_{\beta}^{*} = V_{\beta+1}^{*-1}\ A_{\beta}^{**}\ V_{\beta}^{*} \qquad (\beta = 1,\dots,w-2)\ ,$$

(629) $\quad V_\beta^* \in \Psi_\alpha((l_1,\dots,l_\beta)) \qquad (\beta = 1,\dots,w-1)$

an equivalent descending chain

(630) $\quad \mathfrak{A}^* = \{A_{w-1}^*,\dots,A_1^*\}$

is defined. Let $V_1^* = E$. By theorem 29 there exists V_2^* such that $A_1^* \in \mathcal{J}_\alpha(l;2,1)$. Now let $1 < \beta \le w-2$. Suppose one has already

(631) $\quad A_\iota^* \in \mathcal{J}_\alpha(l;\iota+1,\iota) \qquad (\iota = 1,\dots,\beta-1)$

and let $V_1^*,\dots,V_\beta^*$ be already fixed. By theorem 29 there exists a $V_{\beta+1}^*$ with $A_\beta^* \in \mathcal{J}_\alpha(l;\beta+1,\beta)$. Hence

(632) $\quad A_\beta^* \in \mathcal{J}_\alpha(l;\beta+1,\beta) \qquad (\beta = 1,\dots,w-2)$.

Applying theorem 29 once more one sees that there is a

(633) $\quad U \in \check{\mathcal{M}}(l)$

with

(634) $\quad U^{-1}A_{w-1}^* \in \mathcal{J}_\alpha(l;w,w^*)$.

Set

(635) $\quad A_{w-1} = A_{w-1}^* V_{w-1}$,

(636) $\quad A_\beta = V_{\beta+1}^{-1} A_\beta^* V_\beta \qquad (\beta = 1,\dots,w-2)$,

(637) $\quad V_\beta \in \Delta_\alpha((l_1,\dots,l_\beta)) \qquad (\beta = 1,\dots,w-1)$.

Like in the proof of theorem 29 one sees the following. First one can choose V_{w-1} such that

(638) $\quad U^{-1}A_{w-1} \in \mathcal{J}^*(l;w,w^*)$.

Then one can successively fix $V_{w-2},V_{w-3},\dots,V_1$ such that

(639) $\quad A_\beta \in \mathcal{J}^*(l;\beta+1,\beta) \qquad (\beta = 1,\dots,w-2)$.

Hence

(640) $$\mathfrak{A} = \{A_{w-1}, A_{w-2}, \dots, A_1\}$$

is a reduced descending chain which is equivalent to $\mathfrak{A}^*$ and consequently to $\mathfrak{A}^{**}$. Therefore each equivalence class contains a reduced descending chain.

Now let us assume that two reduced descending chains

(641) $$\mathfrak{A} = \{U\ A_{w-1}, A_{w-2}, \dots, A_1\},$$

(642) $$\mathfrak{A}^* = \{U^* A^*_{w-1}, A^*_{w-2}, \dots, A^*_1\}$$

with

(643) $$U, U^* \in \check{\mathscr{W}}(l),$$

(644) $$A_\beta, A^*_\beta \in \mathscr{J}^*(l; \beta+1, \beta) \qquad (\beta = 1, \dots, w-1),$$

are equivalent. Then

(645) $$A^*_\beta = V^{-1}_{\beta+1}\ A_\beta\ V_\beta \qquad (\beta = 1, \dots, w-1),$$

(646) $$V_\beta \in \Psi_\alpha((l_1, \dots, l_\beta)) \qquad (\beta = 1, \dots, w)$$

(647) $$V_w = U^{*-1}\ U\ .$$

Obviously $V_1 \in \Psi_\alpha(l_1) = \Delta_\alpha(l_1)$. Now let $\beta \geq 1$ and assume $V_\beta \in \Delta_\alpha((l_1, \dots, l_\beta))$. Then $A_\beta V_\beta \in \mathscr{J}_\alpha(l; \beta+1, \beta)$ and because of (644) (645) it follows $V_{\beta+1} \in \Delta_\alpha((l_1, \dots, l_{\beta+1}))$. Hence

(648) $$V_\beta \in \Delta_\alpha((l_1, \dots, l_\beta)) \qquad (\beta=1, \dots, w;\ \alpha=1,2).$$

Now $\check{\mathscr{W}}(l) \subset \Psi_2(l)$ and because of (643), (647), (648) we have $V_w \in \Delta_2(l)$, hence by (643)

(649) $$U^* = U$$

and $V_w = E$. Assume $\beta < w$ and $V_{\beta+1} = E$. From (644), (645) follows

$V_\beta = E$. Hence $V_\beta = E$ $(\beta = 1,\ldots,w)$.

Therefore

(650) $$A_\beta^* = A_\beta \qquad (\beta = 1,\ldots,w) .$$

Hence each equivalence class contains exactly one reduced descending chain.

Let U run over $\check{\mathcal{W}}(1)$ and $\{A_w, A_{w-1},\ldots,A_1\}$ run independently over $\mathcal{O}(1)$. Then $\{UA_w, A_{w-1},\ldots,A_1\}$ runs exactly once over $\mathcal{R}(1)$. This gives the formula (622). Theorem 57 is proved.

§ 9. CHARACTERS

In this paragraph we consider systems of Dirichlet characters. Their theory is developed to that extend that is needed for Selberg's L-series.

DEFINITION 7: Let $\chi_1,\ldots,\chi_{w-1}$ be even characters mod q and form the characterrow

(651) $$\chi = (\chi_1,\ldots,\chi_{w-1})$$

and the $d(w-1) = \frac{w(w-1)}{2}$ products

(652) $$\mathring{\chi}_{\nu\mu} = \chi_\mu \cdots \chi_\nu \qquad (1 \leq \mu \leq \nu \leq w-1) .$$

χ is called "primitive" if all $\mathring{\chi}_{\nu\mu}$ $(1 \leq \mu \leq \nu \leq w-1)$ are primitive.

Theorem 23 tells us under which circumstances primitive characterrows do exist. Choose a row

(653) $$\psi = (\psi_1,\ldots,\psi_w)$$

of even characters such that

(654) $$\psi_{\nu+1}^{-1}\,\psi_\nu = \chi_\nu \qquad (\nu = 1,\dots,w-1).$$

Then ψ is determined by χ up to an even character as common factor. ψ is called "primitive" if and only if χ is primitive, i. e., if all products

(655) $$\mathring{\chi}_{\nu\mu} = \psi_{\nu+1}^{-1}\,\psi_\mu \qquad (1 \leq \mu \leq \nu \leq w-1)$$

are primitive.

Set

(656) $$\tilde{\psi}_\nu = \psi_{w+1-\nu}^{-1} \qquad (\nu = 1,\dots,w),$$

(657) $$\tilde{\psi} = (\tilde{\psi}_1,\dots,\tilde{\psi}_w) \;;$$

(658) $$\check{\psi}_\nu = \psi_{w-\nu}^{-1}(\nu=1,\dots,w-1);\ \check{\psi}_w = \psi_w^{-1} \;;$$

(659) $$\check{\psi} = (\check{\psi}_1,\dots,\check{\psi}_w);$$

(660) $$\psi^* = (\psi_1,\dots,\psi_{w^*}) \;;$$

Then

(661) $$\check{\psi} = (\tilde{\psi}^*,\psi_w^{-1}) \,,$$

(662) $$\tilde{\tilde{\psi}} = \check{\check{\psi}} = \psi \,.$$

Let

(663) $$\mathfrak{z}\psi = (\psi_w,\psi_1,\dots,\psi_{w-1}) \;;$$

(664) $$\mathfrak{z}^*\psi^* = (\psi_{w-1},\psi_1,\dots,\psi_{w-2}) \;;$$

(665) $$\hat{\psi}_1 = \psi_{w^*},\hat{\psi}_\nu = \psi_{\nu-1}\ (\nu=2,\dots,w^*),\hat{\psi}_w = \psi_w \;;$$

(666) $$\hat{\psi} = (\hat{\psi}_1,\dots,\hat{\psi}_w) \,.$$

Then

(667) $$\hat{\psi} = (\mathfrak{z}^*\psi^*, \psi_w) .$$

If one furthermore puts

(668) $$\eta_\alpha(1,\psi) = \tau_\alpha(1^*, \psi_w^{-1}\psi^*) \qquad (\alpha = 1,2) ,$$

one has

(669) $$\eta_\alpha(1,\psi)\eta_\alpha(\check{1},\check{\psi}) = 1 \qquad (\alpha = 1,2) .$$

THEOREM 58: Let χ be a characterrow and $\mathfrak{A}$ be a descending chain. Then the expression

(670) $$\varphi(1,\chi,\mathfrak{A}) = \prod_{\nu=1}^{w-1} \chi_\nu(\hat{h}_\nu(\mathfrak{A}))$$

depends only on the equivalence class of $\mathfrak{A}$. Furthermore

(671) $$\varphi(1,\chi,\mathfrak{A}) = (\prod_{\nu=1}^{w-1} \chi_\nu(\check{h}_\nu \prod_{\iota=1}^{\nu} \prod_{\varkappa=\nu}^{w-1} h_{\varkappa\iota})) ,$$

(672) $$\varphi(1,\chi,\mathfrak{A}) = (\prod_{\iota=1}^{w-1} \mathring{\chi}_{w-1,\iota}(h_\iota^*))(\prod_{1\leq\iota\leq\varkappa\leq w-1} \mathring{\chi}_{\varkappa\iota}(h_{\varkappa\iota})) .$$

PROOF: Since $\hat{h}_\nu(\mathfrak{A})$ $(\nu = 1,\ldots,w-1)$ are equivalence-invariant the same is true for $\varphi(1,\chi,\mathfrak{A})$. From (600), (604) it follows

(673) $$\hat{h}_\nu = \check{h}_\nu \prod_{\iota=1}^{\nu} \prod_{\varkappa=\nu}^{w-1} h_{\varkappa\iota}$$

and hence (671). From (601), (671) one obtains

$$\varphi(1,\chi,\mathfrak{A}) = \prod_{\nu=1}^{w-1} \chi_\nu(\check{h}_\nu \prod_{\iota=1}^{\nu} \prod_{\varkappa=\nu}^{w-1} h_{\varkappa\iota}) = \prod_{\nu=1}^{w-1} \chi_\nu(\prod_{\iota=1}^{\nu}(h_\iota^* \prod_{\varkappa=\nu}^{w-1} h_{\varkappa\iota})) =$$

$$(\prod_{1\leq\iota\leq\nu\leq w-1} \chi_\nu(h_\iota^*))(\prod_{1\leq\iota\leq\nu\leq\varkappa\leq w-1} \chi_\nu(h_{\varkappa\iota})) .$$

Now (672) follows from (652). Theorem 58 is proved.

Let $U \in \Psi_\alpha(1)$ $(\alpha = 1,2)$ and set

$$(674) \qquad \rho(1,\psi,U) = \prod_{\nu=1}^{w} \psi_\nu(\operatorname{Det} \overset{1}{U}_\nu) \quad .$$

THEOREM 59: It is

$$(675) \qquad \rho(1,\psi,U) = \prod_{\nu=1}^{w-1} \mathring{\chi}_{w-1,\nu}(\operatorname{Det} \overset{1}{U}_\nu) \qquad (U \in \Psi_\alpha(1)) ,$$

$$(676) \qquad \rho(\tilde{1},\tilde{\psi},\tilde{U}) = \rho(1,\psi,U) \qquad (U \in \Psi_\alpha(1))(\alpha = 1,2).$$

PROOF: From (250) we deduce

$$\rho(1,\psi,U) = \Big(\prod_{\nu=1}^{w-1} \psi_\nu(\operatorname{Det} \overset{1}{U}_\nu)\Big)\cdot\psi_w^{-1}\Big(\prod_{\nu=1}^{w-1} \operatorname{Det} \overset{1}{U}_\nu\Big)$$
$$= \prod_{\nu=1}^{w-1}(\psi_w^{-1}\psi_\nu)(\operatorname{Det} \overset{1}{U}_\nu) = \prod_{\nu=1}^{w-1} \mathring{\chi}_{w-1,\nu}(\operatorname{Det} \overset{1}{U}_\nu).$$

This gives (675). Formula (676) follows from (223), (311), (656), (674). Theorem 59 is proved.

THEOREM 60: Let

$$(677) \qquad \mathfrak{A} = \{A_{w-1},\dots,A_1\}$$

be a descending chain,

$$(678) \qquad A_\beta = (\overset{1}{A}_{\beta,\nu\mu}) = (a_{\beta,\iota\varkappa}) \qquad (\beta = 1,\dots,w-1) ,$$

where $a_{\beta,\iota\varkappa}$ are the scalar elements of A_β. Then

(679) $$\varphi(1,\chi,\mathfrak{A}) = \prod_{\nu=1}^{w-1} \chi_\nu\Big(\prod_{\iota=1}^{\nu}\prod_{\beta=\nu}^{w-1}(\mathrm{Det}\ \overset{1}{A}_{\beta,\iota})\Big),$$

(680) $$\varphi(1,\chi,\mathfrak{A}) = \prod_{1\le\iota\le\beta\le w-1} \overset{\circ}{\chi}_{\beta\iota}(\mathrm{Det}\ \overset{1}{A}_{\beta\iota}),$$

(681) $$\varphi(1,\chi,\mathfrak{A}) = \prod_{\nu=1}^{w-1} \chi_\nu\Big(\prod_{\iota=1}^{\nu}\prod_{\beta=\nu}^{w-1}\prod_{\varkappa=0}^{1_\iota-1} a_{\beta,k_\iota-\varkappa}\Big),$$

(682) $$\varphi(1,\chi,\mathfrak{A}) = \prod_{1\le\iota\le\beta\le w-1}\ \prod_{\varkappa=0}^{1_\iota-1} \overset{\circ}{\chi}_{\beta\iota}(a_{\beta,k_\iota-\varkappa}).$$

PROOF: From (599), (606) follows

(683) $$\hat{h}_\gamma \equiv \pm \prod_{\mu=1}^{\gamma}\prod_{\nu=\gamma}^{w-1} \mathrm{Det}\ \overset{1}{A}_{\nu\mu} \bmod q \qquad (\gamma = 1,\dots,w-1).$$

From (670), (683) follows (679). This gives

$$\varphi(1,\chi,\mathfrak{A}) = \prod_{1\le\iota\le\nu\le\beta\le w-1} \chi_\nu(\mathrm{Det}\ \overset{1}{A}_{\beta,\iota}) = \prod_{1\le\iota\le\beta\le w-1} \overset{\circ}{\chi}_{\beta\iota}(\mathrm{Det}\ \overset{1}{A}_{\beta\iota}).$$

Herewith one has (680). From (679), (680) one immediately gets (681), (682). Theorem 60 is proved.

§ 10. SELBERG'S ZETA AND L-SERIES

In this paragraph Selberg's zeta- and L-series are defined and the convergence is investigated. Some elementary relations and functional equations are proved.

DEFINITION 8: Let

(684) $$u = (u_1,\dots,u_{w-1})$$

be a row of complex variables, χ a characterrow and

(685) $$Y \in \mathcal{Y}(n) .$$

Define "Selberg's L-series" by

(686) $$\hat{\zeta}(1,\chi,Y,u) = \sum_{\{\mathfrak{A}\}} \varphi(1,\chi,\mathfrak{A}) \prod_{\beta=1}^{w-1} (\mathrm{Det}(Y[B_{w-1,\beta}]))^{-u_\beta} ,$$

(687) $$\hat{\zeta}^*(1,\chi,Y,u) = \sum_{U \in \breve{\mathcal{U}}(1)} \prod_{\beta=1}^{w-1} ((\mathring{\chi}_{w-1,\beta}(\mathrm{Det}\ \overset{1}{U}_\beta))(\mathrm{Det}((Y[U])_{k_\beta}))^{-u_\beta}).$$

In (686) the $n \times k_\beta$ matrices $B_{w-1,\beta}$ are connected with the descending chain $\mathfrak{A}$ by (589). In (686) we sum over all classes of equivalent descending chains of type $\alpha = 1,2$. But by theorem 57 the right-hand-side of (686) does not depend on α. In the right hand side of (687) the set $\breve{\mathcal{U}}(1)$ represents $\Psi_\alpha(1)/\Delta_\alpha(1)$, but also in this case the right-hand-side of (687) does not depend on α. As remarked earlier, $(Y[U])_{k_\beta}$ means the left upper $k_\beta \times k_\beta$ submatrix of $Y[U]$. In case $q = 1$ the Selberg's L-series is also called a "Selberg's zetafunction".

DEFINITION 9: In $\mathbb{R}^{w-1}$ let $\mathfrak{M}(1)$ be the set of all real rows

(688) $$v = (v_1,\ldots,v_{w-1})$$

with

(689) $$v_\nu > \tfrac{1}{2}(l_{\nu+1} + l_\nu) = \tfrac{1}{2}(k_{\nu+1} - k_{\nu-1}) \qquad (\nu = 1,\ldots,w-1).$$

Let

(690) $$u = (u_1,\ldots,u_{w-1})$$

be a complex row. For $\mathfrak{M} \subset \mathbb{R}^{w-1}$ let

(691) $$\mathfrak{M} + i\,\mathbb{R}^{w-1} = \{u \in \mathbb{C}^{w-1};\ \mathrm{Re}\ u \in \mathfrak{M}\} .$$

Set

(692) $$\mathcal{U}(1) = \mathcal{H}(1) + i\,\mathbb{R}^{w-1} .$$

THEOREM 61: In $\mathcal{U}(1)$ the series (687) converges absolutely and represents a holomorphic function in u. Let $\epsilon > 0$ and $\mathcal{M}$ a compact subset of $\mathcal{H}(1)$. For

(693) $$u \in \mathcal{M} + i\,\mathbb{R}^{w-1} ,$$

(694) $$Y \geq \epsilon E$$

the series (687) converges absolutely uniformly with respect to u and Y.

PROOF: It suffices to show that the series (687) converges absolutely uniformly in (693), (694), Set

(695) $$v = (v_1,\ldots,v_{w-1}) = \operatorname{Re} u \quad .$$

Then the "absolut series" belonging to (687) is majorized by

(696) $$\hat{\zeta}^*(1,1,\epsilon E,v) = \epsilon^{-\sum_{\beta=1}^{w-1} k_\beta v_\beta}\, \hat{\zeta}^*(1,1,E,v) \ .$$

Because of $v \in \mathcal{M}$ the expression $\sum_{\beta=1}^{w-1} k_\beta\, v_\beta$ is bounded. Hence it suffices to show that

(697) $$\hat{\zeta}^*(1,1,E,v) = \sum_{U \in \mathcal{W}(1)} \hat{f}(1,U'U,v,0)$$

convergences uniformly for $v \in \mathcal{M}$. Here $\hat{f}$ is defined by (455).

Let

(698) $$b = (b_1,\ldots,b_{w-1}) \in \mathcal{H}(1)$$

be another real row and

(699) $$b_\nu \leq v_\nu \qquad (v \in \mathcal{M})\ (\nu = 1,\ldots,w-1) \ .$$

Since U is integral we have

$$\mathrm{Det}((U'U)_{k_\nu}) \geq 1 \qquad (\nu = 1,\dots,w-1) . \tag{700}$$

Hence

$$\hat{f}(1,U'U,v,0) \leq \hat{f}(1,U'U,b,0) \qquad (v \in \mathfrak{M}) . \tag{701}$$

Therefore the series (697) is majorized by

$$\hat{\zeta}^*(1,1,E,b) . \tag{702}$$

Hence it suffices to prove the convergence of (702) for

$$b \in \mathfrak{M}(1) . \tag{703}$$

We may confine to the case $q = 1$, because this series majorizes the series of the case $q > 1$ since it has more summands. We proceed like in Maaß [33], § 10.

Like in § 6 let $\rho(Y,Y^*)$ be the invariant metric in $\mathcal{Y}(n)$ defined with the help of (392). Let ρ_0 be a positive number and define the ball

$$\vartheta = (Y \in \mathcal{Y}(n) \mid \rho(Y,E) \leq \rho_0) . \tag{704}$$

Here and in the rest of the proof let $d_1, d_2, \dots$ denote constants ≥ 1 which depend on $1, \rho_0, b$.

Let A be a real non singular $n\times n$ matrix and $1 \leq \nu \leq n$. We show the existence of a d_1 with

$$\frac{1}{d_1} \leq \frac{\mathrm{Det}((A'A)_\nu)}{\mathrm{Det}((Y[A])_\nu} \leq d_1 \qquad (Y \in \vartheta) . \tag{705}$$

Let $(A'A)_\nu = C'C$ with a non-singular $\nu\times\nu$ matrix C. Then

$$\frac{\mathrm{Det}((A'A)_\nu)}{\mathrm{Det}((Y[A])_\nu)} = \frac{1}{\mathrm{Det}(((Y[A])_\nu)[C^{-1}])} . \tag{706}$$

From the invariance of ρ and theorem 37 we get

$$\rho(((Y[A]))_\nu)[C^{-1}],E) = \rho((Y[A])_\nu,(A'A)_\nu) \le \rho(Y[A],A'A) = \rho(Y,E) \le \rho_0 \ .$$

Hence (705) follows from (706).

From (455), (705) we deduce

(707) $$\hat{f}(1,A'A,b,0) \le d_2\ \hat{f}(1,Y[A],b,0) \qquad (Y \in \vartheta)\ .$$

Hence

$$\hat{f}(1,A'A,b,0) \int_\vartheta dv_Y \le d_2 \int_\vartheta \hat{f}(1,Y[A],b,0)dv_Y \ .$$

This shows

(708) $$\hat{f}(1,A'A,b,0) \le d_3 \int_\vartheta \hat{f}(1,Y[A],b,0)dv_Y \ .$$

From (697), (708) we see that $\hat{\zeta}^*(1,1,E,b)$ is up to d_3 majorized by

(709) $$\sum_{U \in \Omega(n)/\Delta(1)} \int_\vartheta \hat{f}(1,Y[U],b,0)dv_Y \ .$$

Apply (705) with $A = U$ and observe (700). Then we see

(710) $$\frac{1}{d_1} \le \mathrm{Det}((Y[U])_\nu) \quad (Y \in \vartheta)\ (\nu = 1,\dots,n)\ .$$

For $\nu = n$ we have $\mathrm{Det}(U'U) = 1$. Hence (705) shows

(711) $$\frac{1}{d_1} \le \mathrm{Det}\ Y[U] \le d_1 \qquad (Y \in \vartheta)\ .$$

From Minkowski's reduction theory we see that

(712) $$\vartheta[U] \cap \vartheta \neq \emptyset$$

for N matrices $U \in \Omega(n)$ at the most. Here $\vartheta[U]$ is the image of ϑ under U. Let $\mathcal{E}$ be a fundamental domain of $\Delta(1)$ in $\mathcal{Y}(n)$. Then we may assume that there are $V_1,\dots,V_N \in \Delta(1)$ with

(713) $$\vartheta[U] \in \bigcup_{\nu=1}^{N} \mathcal{E}[V_\nu] \qquad (U \in \Omega(n)/\Delta(1)) .$$

Then (709) is equal to

$$\sum_{U \in \Omega(n)/\Delta(1)} \int_{\vartheta[U]} \hat{f}(1,Y,b,0)dv_Y \leq N\, G_1$$

with

(714) $$G_1 = \int_{\mathcal{E}} \hat{f}(1,Y,b,0)\, dv_Y .$$

$$\frac{1}{d_1} \leq \mathrm{Det}(Y_{k_\nu}) \qquad (\nu = 1,\dots,w-1)$$

$$\frac{1}{d_1} \leq \mathrm{Det}\, Y \leq d_1$$

Therefore it suffices to prove the convergence of (714).

Perform the generalized Jacobitransformation (379), (380), (381) i. e.,

(715) $$Y = R[D] ,$$

(716) $$R = \begin{pmatrix} R_1 & & 0 \\ & \ddots & \\ 0 & & R_w \end{pmatrix} ,$$

(717) $$D = \begin{pmatrix} E & D_{12} & \cdots & D_{1w} \\ & \ddots & & \vdots \\ 0 & & \ddots & E \end{pmatrix} .$$

Then by (385)

(718) $$dv_Y = \prod_{\nu=1}^{w} \{(\mathrm{Det}\, R_\nu)^{\frac{1}{2}(n-k_\nu-k_{\nu-1})} dv_{R_\nu}\}[dD] .$$

The fundamentaldomain $\mathcal{E}$ of $\Delta(1)$ may be described by $R_\nu \in \mathfrak{M}(l_\nu)$ $(\nu = 1,\dots,w)$ and the condition that all elements of D are between 0 and 1. Substituting this into (714) and immediately integrating over D gives

$$(719) \qquad G_1 = \int \hat{f}(1,R,b,0)\Big(\prod_{\nu=1}^{w}(\operatorname{Det} R_\nu)^{\frac{1}{2}(n-k_\nu-k_{\nu-1})}\Big)dv_{R_1}\dots dv_{R_w}.$$

$$R_\nu \in \mathfrak{M}(l_\nu) \quad (\nu = 1,\dots,w)$$

$$\frac{1}{d_1} \le \prod_{\mu=1}^{\nu}\operatorname{Det} R_\mu \qquad (\nu = 1,\dots,w-1)$$

$$\frac{1}{d_1} \le \prod_{\mu=1}^{w}\operatorname{Det} R_\mu \le d_1$$

From (455) we get

$$(720) \qquad \hat{f}(1,R,v,0) = \prod_{\nu=1}^{w-1}(\operatorname{Det} R_\nu)^{-b_\nu-\dots-b_{w-1}} \quad .$$

Set

$$(721) \qquad g_\nu = \frac{1}{2}(k_\nu + k_{\nu-1} - n) + b_\nu + \dots + b_{w-1} \qquad (\nu = 1,\dots,w-1),$$

$$(722) \qquad g_w = \frac{1}{2}\, k_{w-1} \; .$$

Then

$$(723) \qquad g_1 > g_2 > \dots > g_w > 0 \quad .$$

Inserting (720), (721), (722) into (719) gives

$$(724) \qquad G_1 = \int_{R_\nu \in \mathfrak{M}(l_\nu)} \Big(\prod_{\nu=1}^{w}(\operatorname{Det} R_\nu)^{-g_\nu}\Big)dv_{R_1}\dots dv_{R_w} \; .$$

$$\frac{1}{d_1} \le \prod_{\mu=1}^{\nu}\operatorname{Det} R_\mu \qquad (\nu = 1,\dots,w-1)$$

$$\frac{1}{d_1} \le \prod_{\mu=1}^{w}\operatorname{Det} R_\mu \le d_1 \quad .$$

Apply (377), i. e.,

(725) $$\int\limits_{\substack{\mathcal{M}(n)\\ y_1 \leq \operatorname{Det} Y \leq y_2}} \Phi(\operatorname{Det} Y)\,dv_Y = \frac{n+1}{2} v_n \int\limits_{y_1}^{y_2} \Phi(y)\,\frac{dy}{y} .$$

From this we deduce

(726) $$\int\limits_{\substack{\mathcal{M}(n)\\ y_1 \leq \operatorname{Det} Y}} (\operatorname{Det} Y)^{-g}\,dv_Y = \frac{n+1}{2g} v_n\, y_1^{-g} \quad (g > 0) .$$

Applying this several times to (724) gives $G_1 = \text{const}\; G_2$ with

(727) $$G_2 = \int\limits_{\substack{\frac{1}{d_1} \leq \prod_{\mu=1}^{\nu} r_\mu \quad (\nu = 1,\dots,w-1)\\ \frac{1}{d_1} \leq \prod_{\mu=1}^{w} r_\mu \leq d_1}} \Big(\prod_{\mu=1}^{w} r_\mu^{-g_\mu - 1} \Big) dr_1 \dots dr_w .$$

Substitute

(728) $$t_\nu = \prod_{\mu=1}^{\nu} r_\mu \qquad (\nu = 1,\dots,w) .$$

Then

$$\frac{\partial(t_1,\dots,t_w)}{\partial(r_1,\dots,r_w)} = \operatorname{Det} \begin{pmatrix} 1 & 0 & \dots & 0 \\ * & t_1 & \ddots & \vdots \\ \vdots & & \ddots & 0 \\ * & \dots & * & t_{w-1} \end{pmatrix} = t_1 \dots t_{w-1} .$$

Hence

$$(729)\qquad \frac{\partial(r_1,\dots,r_w)}{\partial(t_1,\dots,t_w)} = t_1^{-1} \cdots t_{w-1}^{-1} .$$

Furthermore

$$(730)\qquad \prod_{\mu=1}^{w} r_\mu^{-1-g_\mu} = \left(\prod_{\nu=1}^{w-1} t_\nu^{g_{\nu+1}-g_\nu}\right) t_w^{-1-g_w} .$$

Therefore

$$(731)\qquad G_2 = \left(\prod_{\nu=1}^{w-1} \int_{\frac{1}{d_1}}^{\infty} t_\nu^{g_{\nu+1}-g_\nu-1}\, dt_\nu\right) \int_{\frac{1}{d_1}}^{d_1} t_w^{-1-g_w}\, dt_w .$$

This converges because of (723). Theorem 61 is proved.

THEOREM 62: In $\mathcal{U}(1)$ the series (686) converges absolutely and represents a holomorphic function in u. Let $\mathfrak{M}$ be a compact subset of $\mathcal{H}(1)$. For

$$(732)\qquad u \in \mathfrak{M} + i\,\mathbb{R}^{w-1}$$

the series (686) converges absolutely uniformly with respect to u. Let $L(\chi,z)$ denote the ordinary Dirichlet L-series for χ and put

$$(733)\qquad \hat{L}(1,\chi,u) = \prod_{1\le\mu\le\nu\le w-1} \prod_{\varkappa=0}^{1_\mu-1} L(\mathring{\chi}_{\nu\mu}, 2(u_\mu+\dots+u_\nu)-k_\nu+k_\mu-\varkappa).$$

Then

$$(734)\qquad \hat{\zeta}(1,\chi,Y,u) = \hat{L}(1,\chi,u)\,\hat{\zeta}^{*}(1,\chi,Y,u) .$$

PROOF: First we prove (734) by a formal computation. Afterwards we shall see that all steps are justified. Then the convergence properties of (686) will follow from theorem 61.

Because of theorem 57 one may in (686) sum over all reduced descending chains

(735) $$\mathfrak{m} = \{UD_{w-1}, D_{w-2}, \dots, D_1\}$$

with

(736) $$U = (\overset{1}{U}_{\nu\mu}) \in \check{\mathfrak{m}}(1),$$

(737) $$D_\beta = (\overset{1}{D}_{\beta,\nu\mu}) = (d_{\beta,\iota\varkappa}) \in \mathcal{J}^*(1;\beta+1,\beta) \quad (\beta = 1,\dots,w-1).$$

Then

(738) $$B_{w-1,\beta} = UD_{w-1}\cdots D_\beta \qquad (\beta = 1,\dots,w-1)\ .$$

Hence

(739) $$\mathrm{Det}(Y[B_{w-1,\beta}]) = \mathrm{Det}((Y[U])_{k_\beta})\prod_{\nu=\beta}^{w-1}\prod_{\mu=1}^{\beta}(\mathrm{Det}\ \overset{1}{D}_{\nu\mu})^2 \quad (\beta=1,\dots,w-1).$$

From (616), (619), (672) we deduce

(740) $$\varphi(1,\chi,\mathfrak{m}) = (\prod_{\nu=1}^{w-1}\mathring{\chi}_{w-1,\nu}(\mathrm{Det}\ \overset{1}{U}_\nu))(\prod_{1\le\mu\le\nu\le w-1}\mathring{\chi}_{\nu\mu}(\mathrm{Det}\ \overset{1}{D}_{\nu\mu})).$$

(741) $$\hat{\zeta}(1,\chi,Y,u) = \sum_{U\in\check{\mathfrak{m}}(1)}(\prod_{\beta=1}^{w-1}(\mathring{\chi}_{w-1,\beta}(\mathrm{Det}\ \overset{1}{U}_\beta))\mathrm{Det}((Y[U])_{k_\beta})^{-u_\beta})\times$$
$$\sum_{\mathcal{G}(1)}((\prod_{1\le\mu\le\nu\le w-1}\mathring{\chi}_{\nu\mu}(\mathrm{Det}\ \overset{1}{D}_{\nu\mu}))(\prod_{1\le\mu\le\beta\le\nu\le w-1}(\mathrm{Det}\ \overset{1}{D}_{\nu\mu})^{-2u_\beta}))\quad .$$

Hence

(742) $$\hat{\zeta}(1,\chi,Y,u) = \Lambda\ \hat{\zeta}^*(1,\chi,Y,u)$$

with

(743) $$\Lambda = \sum_{\mathcal{G}(1)}(\prod_{1\le\mu\le\nu\le w-1}\mathring{\chi}_{\nu\mu}(\mathrm{Det}\ \overset{1}{D}_{\nu\mu})(\mathrm{Det}\ \overset{1}{D}_{\nu\mu})^{-2(u_\mu+\dots+u_\nu)})\quad .$$

Formula (743) gives

$$(744)\qquad \Lambda = \sum_{\mathcal{G}(1)} \Big(\prod_{1 \le \mu \le \nu \le w-1} \prod_{\varkappa=0}^{l_\mu - 1} \big(\mathring{\chi}_{\nu\mu}(d_{\nu,k_\mu-\varkappa})\, d_{\nu,k_\mu-\varkappa}^{-2(u_\mu+\dots+u_\nu)}\big) \Big) .$$

Hence because of theorem 26

$$(745)\qquad \Lambda = \prod_{1\le\mu\le\nu\le w-1} \prod_{\varkappa=0}^{l_\mu-1} \sum_{d_{\nu,k_\mu-\varkappa}=1}^{\infty} \mathring{\chi}_{\nu\mu}(d_{\nu,k_\mu-\varkappa})\, d_{\nu,k_\mu-\varkappa}^{-2(u_\mu+\dots+u_\nu)-k_\mu+k_\nu+\varkappa}$$

and because of (733)

$$(746)\qquad \Lambda = \hat{L}(1,\chi,u) .$$

From (742), (746) follows (734) provided that all substitutions are allowed.

We shall prove that all substitutions are allowed for $u \in \mathcal{U}(1)$. Let χ_o be the principal character mod q and

$$(747)\qquad \chi_p = (\chi_o,\dots,\chi_o)$$

with w-1 times χ_o. Let $v \in \mathcal{V}(1)$. From wellknown properties of Dirichlet's L-series (see for instance Landau [22]) and from theorem 61 it follows that all series on the right-hand-side of (734) converge absolutely for $u \in \mathcal{U}(1)$ and that they converge absolutely uniformly for (732). Now let $\chi = \chi_p$ and $u = v \in \mathcal{V}(1)$. Then all terms in the preceding computation are ≥ 0. Hence (734) must be true for $u = v$, $\chi = \chi_p$. If $u \in \mathcal{U}(1)$ and χ are arbitrary the absolute values of all terms are majorized by the terms of the case Re u, χ_p. Hence (734) must hold for all $u \in \mathcal{U}(1)$ and all χ. Then the convergence properties of $\hat{\zeta}(1,\chi,Y,u)$ follow from theorem 61. Theorem 62 is proved.

THEOREM 63: The functions (686), (687) are homogeneous in Y of degree

$$(748)\qquad -\sum_{\mu=1}^{w-1} k_\mu u_\mu .$$

For $U \in \Psi_\alpha(1)$ $(\alpha = 1,2)$ we have

(749) $\hat{\zeta}^{*}(1,\chi,Y[U],u)=(\prod_{\mu=1}^{w-1} \overset{\circ}{\chi}_{w-1,\mu}(\mathrm{Det}\ \overset{1}{U}_{\mu}))^{-1}\ \hat{\zeta}^{*}(1,\chi,Y,u)$,

(750) $\hat{\zeta}(1,\chi,Y[U],u)=(\prod_{\mu=1}^{w-1} \overset{\circ}{\chi}_{w-1,\mu}(\mathrm{Det}\ \overset{1}{U}_{\mu}))^{-1}\ \hat{\zeta}(1,\chi,Y,u)$.

PROOF: It suffices to prove (749). It is $\mathscr{N}(1) = \Psi_{\alpha}(1)/\Delta_{\alpha}(1)$ ($\alpha = 1,2$). Hence (687) may be written

(751) $\hat{\zeta}^{*}(1,\chi,Y,u)= \sum_{V \in \Psi_{\alpha}(1)/\Delta_{\alpha}(1)} \prod_{\beta=1}^{w-1}((\overset{\circ}{\chi}_{w-1,\beta}(\mathrm{Det}\ \overset{1}{V}_{\beta}))(\mathrm{Det}((Y[UV])_{k_{\beta}}))^{-u_{\beta}})$.

With V also UV runs over $\Psi_{\alpha}(1)/\Delta_{\alpha}(1)$. Hence

$$\hat{\zeta}^{*}(1,\chi,Y,u)= \sum_{V \in \Psi_{\alpha}(1)/\Delta_{\alpha}(1)} \prod_{\beta=1}^{w-1}((\overset{\circ}{\chi}_{w-1,\beta}(\mathrm{Det}\ \overset{1}{U}_{\beta}\mathrm{Det}\ \overset{1}{V}_{\beta}))(\mathrm{Det}((Y[UV])_{k_{\beta}}))^{-u_{\beta}})$$

$$= \prod_{\beta=1}^{w-1} (\overset{\circ}{\chi}_{w-1,\beta}(\mathrm{Det}\ \overset{1}{U}_{\beta}))\ \hat{\zeta}^{*}(1,\chi,Y[U],u) .$$

This proves (749). Formula (750) follows then from (734). Theorem 63 is proved.

Let the variable row $s = (s_1,\ldots,s_w)$ and u be connected by (452). Let ψ and χ be connected by (654). Set

(752) $$\zeta(1,\psi,Y,s) = \hat{\zeta}(1,\chi,Y,u),$$

(753) $$\zeta^{*}(1,\psi,Y,s) = \hat{\zeta}^{*}(1,\chi,Y,u) ,$$

(754) $$L(1,\psi,s) = \hat{L}(1,\chi,u) .$$

Then by (734)

(755) $$\zeta(1,\psi,Y,s) = L(1,\psi,s)\zeta^{*}(1,\psi,Y,s) .$$

Furthermore because of (655), (680)

(756) $\quad \zeta(l,\psi,Y,s) =$

$$\sum_{\{\alpha\}} \Big(\prod_{1 \le \iota \le \beta \le w-1} (\psi_{\beta+1}^{-1}\psi_\iota)(\text{Det } A_{\beta\iota}^{l}) \Big) \Big(\prod_{\nu=1}^{w-1} (\text{Det}(Y[B_{w-1,\nu}]))^{s_\nu - s_{\nu+1} - \frac{l_{\nu+1}+l_\nu}{4}} \Big) ;$$

(757) $\quad \zeta^*(l,\psi,Y,s) =$

$$\sum_{U \in \Psi_\alpha(l)/\Delta_\alpha(l)} \rho(l,\psi,U) \Big(\prod_{\nu=1}^{w-1} (\text{Det}((Y[U])_{k_\nu}))^{s_\nu - s_{\nu+1} - \frac{l_{\nu+1}+l_\nu}{4}} ;$$

(758) $$L(l,\psi,s) = \prod_{1 \le \mu \le \nu \le w-1} \prod_{\varkappa=0}^{l_\mu - 1} L(\psi_{\nu+1}^{-1}\psi_\mu; 2(s_{\nu+1}-s_\mu) + \tfrac{1}{2}(l_{\nu+1} + l_\mu) - \varkappa) .$$

THEOREM 64: The functions (756), (757) are homogeneous in Y of degree

(759) $$n(\{l,s\} - s_w - \tfrac{1}{4} k_{w-1}) .$$

Furthermore for $U \in \Psi_\alpha(l)$; $\alpha = 1,2$

(760) $$\zeta^*(l,\psi,Y[U],s) = \rho^{-1}(l,\psi,U)\zeta^*(l,\psi,Y,s) ,$$

(761) $$\zeta(l,\psi,Y[U],s) = \rho^{-1}(l,\psi,U)\zeta(l,\psi,Y,s) .$$

PROOF: Apply theorem 63. One has

$$-\sum_{\mu=1}^{w-1} k_\mu u_\mu = -\sum_{\mu=1}^{w-1} k_\mu s_{\mu+1} + \sum_{\mu=1}^{w-1} k_\mu s_\mu$$

$$-\frac{1}{4}\sum_{\mu=1}^{w-1} k_\mu(l_{\mu+1}+l_\mu) = -\sum_{\mu=1}^{w} k_{\mu-1}s_\mu + \sum_{\mu=1}^{w} k_\mu s_\mu - k_w s_w - \frac{1}{4}\sum_{\mu=1}^{w-1} k_\mu(k_{\mu+1}-k_{\mu-1}) =$$

$$\sum_{\mu=1}^{w} l_\mu s_\mu - ns_w - \frac{1}{4}\sum_{\mu=1}^{w-1} k_\mu k_{\mu+1} + \frac{1}{4}\sum_{\mu=1}^{w-1} k_{\mu-1}\, k_\mu =$$

$$n\{l,s\} - ns_w - \frac{1}{4}\sum_{\mu=1}^{w-1} k_\mu\, k_{\mu+1} + \frac{1}{4}\sum_{\mu=o}^{w-2} k_\mu\, k_{\mu+1} =$$

$$n(\{l,s\} - s_w - \frac{1}{4}k_{w-1}) \; .$$

This proves (759). Formulas (760), (761) follow from (675), (749), (750). Theorem 64 is proved.

THEOREM 65: Set

$$\Lambda(l,\psi,Y,s) = (\operatorname{Det} Y)^{s_w + \frac{n^*}{4}}\,\zeta^*(l,\psi,Y,s) \; . \tag{762}$$

The series (756), (757), (762) converge absolutely for

$$\sigma_{\nu+1} - \sigma_\nu > \frac{l_{\nu+1} + l_\nu}{4} \qquad (\nu = 1,\ldots,w-1) \tag{763}$$

and are there holomorphic. The domain (763) is invariant under the substitution

$$l \to \tilde{l} \; , \quad s \to \tilde{s} \; . \tag{764}$$

The function (762) is homogeneous in Y of degree

$$n\{l,s\}. \tag{765}$$

One has

(766) $$\Lambda(1,\psi,Y,s) = \sum_{U \in \Psi_\alpha(1)/\Delta_\alpha(1)} \rho(1,\psi,U)f(1,Y[U],s) ,$$

(767) $$\Lambda(1,\psi,Y[U],s) = \rho^{-1}(1,\psi,U)\Lambda(1,\psi,Y,s) \quad (U \in \Psi_\alpha(1)),$$

(768) $$\Lambda(\tilde{1},\tilde{\psi},\tilde{Y},\tilde{s}) = \Lambda(1,\psi,Y,s) ,$$

(769) $$(\mathrm{Det}\ Y)^{s_w + \frac{n^*}{4}} \zeta(1,\psi,Y,s) = L(1,\psi,s)\Lambda(1,\psi,Y,s) .$$

PROOF: The assertion about convergence and holomorphy follows from theorems 61, 62. The invariance of the domain (763) under the substitution (764) is easily computed. (766) follows easily from (424), (757), (762). The degree (765) follows from theorem 38 and also from theorem 64. Formula (767) follows from (760), (762). Formula (769) is a consequence of (755) and (762). Finally (768) follows from (310), (433), (676), (766). Theorem 65 is proved.

§ 11. ANALYTIC CONTINUATION

In this paragraph the analytic continuation of Selberg's zeta- and L-series is proved and an important functional equation is derived. The method is similar to Maaβ [33], § 17. From the functional equation we derive a result about Gaussian sums. We show that Selberg's L-series are holomorphic in certain domains.

THEOREM 66: Let $\Gamma(\ldots)$ be the Γ-function, s a complex variable, χ a even character mod q and

(770) $$\xi(\chi,s) = \left(\frac{\pi}{q}\right)^{-s}\Gamma(s)L(\chi,2s).$$

For $q > 1$ the functions $\xi(\chi,s)$ and $L(\chi,2s)$ are holomorphic in $\mathbb{C}$. For $q = 1$ the functions $s(\frac{1}{2} - s)\xi(\chi,s)$ and $(\frac{1}{2} - s)L(\chi,2s)$ are holomorphic in $\mathbb{C}$. Let χ be primitive. Then all zeros of

$\xi(\chi,s)$ lie in the strip

(771) $$0 < \mathrm{Re}\ s < \tfrac{1}{2}\ .$$

PROOF: See Landau [22], § 128.

THEOREM 67: Let $a \in \mathbb{C}$,

(772) $$Y \in \mathcal{Y}(n),\ T \in \mathcal{Y}(n^*)\ .$$

Then the functions

(773) $$\theta_\alpha(l,\psi_2^{-1}\psi_1,Y,T)(\mathrm{Det}\ T)^a \qquad (w = 2;\ \alpha = 1,2)\ ,$$

(774) $$\theta_\alpha(l,\psi_w^{-1}\psi^*,Y,T)(\mathrm{Det}\ T)^a\ \Lambda(\widetilde{l^*},\widetilde{\psi^*},T,\widetilde{s^*})\ (w \geq 3;\ \alpha = 1,2)$$

is invariant under the substitution

(775) $$T \rightarrow T[U] \qquad (U \in \Psi_\alpha(l);\ \alpha = 1,2).$$

PROOF: Apply theorem 46 and formula (767).

Define

(776) $$\check{\xi}(l,\psi,s) = q^{\frac{n^*(n^*-1)}{4}} \prod_{\nu=1}^{w^*} \prod_{\iota=0}^{l_\nu -1} \xi(\psi_w^{-1}\psi_\nu, s_w - s_\nu + \frac{l_w + l_\nu}{4} - \frac{1}{2})\ ,$$

(777) $$\delta(l,s) = \prod_{\nu=1}^{w^*} \prod_{\iota=0}^{l_\nu -1} \{(s_\nu - s_w + \frac{l_w+l_\nu}{4} - \frac{1}{2})(s_w - s_\nu + \frac{l_w+l_\nu}{4} - \frac{1}{2})\},$$

(778) $$\xi(l,\psi,s) = \begin{cases} \delta(l,s)\check{\xi}(l,\psi,s) & (q = 1) \\ \check{\xi}(l,\psi,s) & (q > 1) \end{cases}\ .$$

THEOREM 68: In the domain

$$\sigma_{\nu+1} - \sigma_\nu > \frac{l_{\nu+1} + l_\nu}{4} \qquad (\nu = 1,\ldots,w-1) \tag{779}$$

one has for $\alpha = 1,2$:

$$\xi(l,\psi,s)(\operatorname{Det} Y)^{-\{l,s\}}\Lambda(l,\psi,Y,s) = \tag{780}$$

$$\tfrac{1}{2}(\operatorname{Det} Y)^{s_2-\{l,s\}} \int\limits_{T \in \mathfrak{F}_\alpha(\widetilde{l^*})} \theta_\alpha(l,\psi_2^{-1}\psi_1,Y,T)(\operatorname{Det} T)^{s_2-s_1} dv_T \qquad (w = 2),$$

$$\xi(l,\psi,s)(\operatorname{Det} Y)^{-\{l,s\}} \Lambda(l,\psi,Y,s) = \tag{781}$$

$$\tfrac{1}{2}(\operatorname{Det} Y)^{s_w-\{l,s\}} \int\limits_{T \in \mathfrak{F}_\alpha(\widetilde{l^*})} \theta_\alpha(l,\psi_w^{-1}\psi^*,Y,T)(\operatorname{Det} T)^{s_w}\Lambda(\widetilde{l^*},\widetilde{\psi^*},T,\widetilde{s^*})dv_T \quad (w \geq 3).$$

PROOF: Because of theorem 67 the integrals in (780), (781) do not depend on the choice of the fundamental domain $\mathfrak{F}_\alpha(\widetilde{l^*})$. Like at the end of the proof of theorem 62 one sees that the following substitutions are allowed.

First let $w = 2$. We put

$$\Xi = \tfrac{1}{2}(\operatorname{Det} Y)^{s_2} \int\limits_{T \in \mathfrak{F}_\alpha(\widetilde{l^*})} \theta_\alpha(l,\psi_2^{-1}\psi_1,Y,T)(\operatorname{Det} T)^{s_2-s_1} dv_T \; . \tag{782}$$

We have to prove

$$\Xi = \xi(l,\psi,s)\Lambda(l,\psi,Y,s) \; . \tag{783}$$

First let $q = 1$. From theorem 49 one deduces

$$\Xi = (\operatorname{Det} Y)^{s_2+\frac{n^*}{4}} \sum_{A \in \mathfrak{D}_1(l;w,w^*)} \times$$

$$\frac{1}{2} \int_{T \in \tilde{\mathfrak{F}}_1(1^*)} (\mathrm{Det}\ T)^{s_2 - s_1 + \frac{n}{4}} P_{\frac{n}{2}}(T) \exp(-\pi\ \mathrm{Tr}(Y[AW(n^*)]T))\ dv_T .$$

Hence by (399), (400), (534)

$$\Xi = (\mathrm{Det}\ Y)^{s_2 + \frac{n^*}{4}} \sum_{A \in \mathfrak{D}_1(1;w,w^*)/\Psi_1(1^*)} \mathrm{Det}(-\pi Y[A]) \quad \times$$

$$\int_{T \in \mathcal{Y}(n^*)} (\mathrm{Det}\ T)^{s_2 - s_1 - \frac{n}{4}} (\hat{M}_{n^*}((\mathrm{Det}\ T)^{\frac{n}{2}+1} \exp(-\pi\ \mathrm{Tr}(Y[AW(n^*)]T)) dv_T .$$

Therefore

(784)
$$\Xi = (\mathrm{Det}\ Y)^{s_2 + \frac{n^*}{4}} \sum_{A \in \mathfrak{D}_1(1;w,w^*)/\Psi_1(1^*)} \mathrm{Det}(-\pi Y[A]) \quad \times$$

$$\int_{T \in \mathcal{Y}(n^*)} (M_{n^*}(\mathrm{Det}\ T)^{s_2 - s_1 - \frac{n}{4}}) (\mathrm{Det}\ T)^{\frac{n}{2}+1} \exp(-\pi\ \mathrm{Tr}(Y[AW(n^*)]T)) dv_T .$$

For $Y \in \mathcal{Y}(n)$ consider the integral

(785)
$$F(n,Y,s) = \int_{V \in \mathcal{Y}(n)} (\mathrm{Det}\ V)^s \exp(-\mathrm{Tr}(YV))\ dv_V ,$$

which is a special case of $J(1,X,s)$ considered in theorem 40. Theorem 40 shows

(786)
$$(\mathrm{Det}\ Y)^s = \frac{F(n,Y,-s)}{F(n,E,-s)}$$

with

(787)
$$F(n,E,s) = \pi^{\frac{n(n-1)}{4}} \prod_{\nu=0}^{n-1} \Gamma(s - \tfrac{\nu}{2}) .$$

Therefore by (399), (534)

$$M_n(Y)(\mathrm{Det}\ Y)^s = F^{-1}(n,E,-s) \int_{V \in \mathfrak{P}(n)} (\mathrm{Det}\ V)^{-s} M_n(Y) \exp(-\mathrm{Tr}(YV))dv_V =$$

$$F^{-1}(n,E,-s)(\mathrm{Det}\ Y)(-1)^n F(n,Y,1-s) = (\mathrm{Det}\ Y)^s (-1)^n \frac{F(n,E,1-s)}{F(n,E,-s)} .$$

Hence

$$(788) \qquad M_n(Y)(\mathrm{Det}\ Y)^s = \left(\prod_{\nu=0}^{n-1}(s + \tfrac{\nu}{2})\right)(\mathrm{Det}\ Y)^s .$$

From (784), (788) we obtain

$$(789) \quad \Xi = (-1)^{n^*} \prod_{\nu=0}^{n^*-1}(s_2 - s_1 - \tfrac{n}{4} + \tfrac{\nu}{2})(\mathrm{Det}\ Y)^{s_2 + \frac{n^*}{4}} \sum_{A \in \mathfrak{B}_1(1;w,w^*)/\Psi_1(1^*)} (\mathrm{Det}(\pi Y[A])) \quad \times$$

$$\int_{T \in \mathfrak{P}(n^*)} (\mathrm{Det}\ T)^{s_2 - s_1 + \frac{n}{4} + 1} \exp(-\pi\ \mathrm{Tr}(Y[AW(n^*)]T))dv_T .$$

From (309), (785), (789) we get

$$(790) \qquad \Xi = \prod_{\nu=0}^{n^*-1}(s_1 - s_2 + \tfrac{n}{4} - \tfrac{\nu}{2})(\mathrm{Det}\ Y)^{s_2 + \frac{n^*}{4}} \quad \times$$

$$\sum_{A \in \mathfrak{B}_1(1;w,w^*)/\Psi_1(1^*)} \mathrm{Det}(\pi Y[A]) F(n^*, \pi Y[AW(n^*)], s_2 - s_1 + \tfrac{n}{4} + 1)$$

and by (786), (787)

$$(791) \qquad \Xi = \pi^{\frac{n^*(n^*-1)}{4}} \prod_{\nu=0}^{n^*-1} \{(s_1 - s_2 + \tfrac{n}{4} - \tfrac{\nu}{2})(s_2 - s_1 + \tfrac{n}{4} - \tfrac{\nu}{2})\} \quad \times$$

$$\prod_{\nu=0}^{n^*-1} \Gamma(s_2 - s_1 + \tfrac{n}{4} - \tfrac{\nu}{2})(\mathrm{Det}\ Y)^{s_2 + \frac{n^*}{4}} \sum_{A \in \mathfrak{B}_1(1;w,w^*)/\Psi_1(1^*)} (\mathrm{Det}(\pi Y[A]))^{s_1 - s_2 - \frac{n}{4}} .$$

From (756) with χ the principal character, (777) and (791) we get

$$(792)\quad \Xi = \pi^{\frac{n^*(n^*-1)}{4} + n^*(s_1-s_2-\frac{n}{4})} \delta(1,s)\left(\prod_{\nu=0}^{n^*-1} \Gamma(s_2-s_1+\frac{n}{4}-\frac{\nu}{2})\right) \times$$

$$(\mathrm{Det}\ Y)^{s_2+\frac{n^*}{4}} \zeta(1,\psi,Y,s)$$

and because of (769)

$$(793)\quad \Xi = \pi^{\frac{n^*(n^*-1)}{4} + n^*(s_1-s_2-\frac{n}{4})} \delta(1,s)\left(\prod_{\nu=0}^{n^*-1} \Gamma(s_2-s_1+\frac{n}{4}-\frac{\nu}{2})\right) \times$$

$$L(1,\psi,s)\Lambda(1,\psi,Y,s)\ .$$

From (758), (770), (776), (778), (793) we get (783). This concludes the case $w = 2$, $q = 1$.

Now let $w = 2, q > 1$. From theorem 43, (529), (531), (782) we get

$$(794)\qquad \Xi = (\mathrm{Det}\ Y)^{s_2+\frac{n^*}{4}} \sum_{A \in \mathcal{B}_\alpha(1;w,w^*)} (\psi_2^{-1}\psi_1)(\mathrm{Det}\ \overset{1}{A_1}) \quad \times$$

$$\frac{1}{2} \int_{T \in \widetilde{\mathcal{J}_\alpha(1^*)}} (\mathrm{Det}\ T)^{s_2-s_1+\frac{n}{4}} \exp(-\frac{\pi}{q}\ \mathrm{Tr}(Y[AW(n^*)]T)\ dv_T\ .$$

Formulas (309), (794) give us

$$(795)\quad \Xi = (\mathrm{Det}\ Y)^{s_2+\frac{n^*}{4}} \sum_{A \in \mathcal{B}_\alpha(1;w,w^*)/\Psi_\alpha(1^*)} (\psi_2^{-1}\psi_1)(\mathrm{Det}\ \overset{1}{A_1})F(n^*,\pi Y[AW(n^*)],s_2-s_1+\frac{n}{4})\ .$$

From (786), (787) we deduce

$$(796)\qquad \Xi = (\mathrm{Det}\ Y)^{s_2+\frac{n^*}{4}}\ \pi^{\frac{n^*(n^*-1)}{4} + n^*(s_1-s_2-\frac{n}{4})} \quad \times$$

$$\left(\prod_{\nu=0}^{n^*-1}\Gamma(s_2-s_1+\tfrac{n}{4}-\tfrac{\nu}{2})\right)\sum_{A\in\mathcal{B}_\alpha(1;w,w^*)/\Psi_\alpha(1^*)}(\psi_2^{-1}\psi_1)(\mathrm{Det}\ \overset{1}{A}_1)(\mathrm{Det}\ Y[A])^{s_1-s_2-\frac{n}{4}} .$$

Now (783) follows like in the case $q = 1$. The case $w = 2$ is complete.

For $w \geq 3$ we set

(797) $$\Xi = \tfrac{1}{2}(\mathrm{Det}\ Y)^{s_w}\int_{T\in\mathcal{F}_\alpha(\widetilde{1^*})}\theta_\alpha(1,\psi_w^{-1}\psi^*,Y,T)(\mathrm{Det}\ T)^{s_w}\Lambda(\widetilde{1^*},\widetilde{\psi^*},T,\widetilde{s^*})dv_T .$$

Again we have to prove (783).

Set

(798) $$S = \begin{cases} P_{\frac{n}{2}} & (q = 1) \\ 1 & (q > 1) \end{cases} .$$

Then from theorems 43, 49 and (531) we deduce

(799) $$\theta_\alpha(1,\chi,Y,T)=(\mathrm{Det}\ Y)^{\frac{n^*}{4}}(\mathrm{Det}\ T)^{\frac{n}{4}}\sum_{A\in\mathcal{B}_\alpha(1;w,w^*)}\chi_1(\mathrm{Det}\ \overset{1}{A}_1)\dots\chi_{w^*}(\mathrm{Det}\ \overset{1}{A}_{w^*})\ \times$$

$$S\exp(-\tfrac{\pi}{q}\,\mathrm{Tr}(Y[AW(n^*)]T)) .$$

Inserting (766), (799) into (797) gives

(800) $$\Xi = (\mathrm{Det}\ Y)^{s_w+\frac{n^*}{4}}\sum_{U\in\Psi_\alpha(\widetilde{1^*})/\Delta_\alpha(\widetilde{1^*})}\ \sum_{A\in\mathcal{B}_\alpha(1;w,w^*)}\ \times$$

$$(\psi_w^{-1}\psi_1)(\mathrm{Det}\ \overset{1}{A}_1)\dots(\psi_w^{-1}\psi_{w^*})(\mathrm{Det}\ \overset{1}{A}_{w^*})\rho(\widetilde{1^*},\widetilde{\psi^*},U)\ \times$$

$$\tfrac{1}{2}\int_{T\in\mathcal{F}_\alpha(\widetilde{1^*})}f(\widetilde{1^*},T[U],\widetilde{s^*}+(s_w+\tfrac{n}{4})\,e(w^*))S\exp(-\tfrac{\pi}{q}\,\mathrm{Tr}(Y[AW(n^*)]T))\,dv_T .$$

Keep $U \in \Psi_\alpha(\widetilde{1^*})/\Delta_\alpha(\widetilde{1^*})$ fixed. Then because of (309) we have $W(n^*)U'W(n^*) \in \Psi_\alpha(1^*_{\alpha})$. Set

$$A = A^*W(n^*)U'W(n^*) \ . \tag{801}$$

Then with A also A^* runs over $\mathcal{B}_\alpha(1;w,w^*)$. From (309), (310), (311), (674) we deduce

$$(\psi_w^{-1}\psi_1)(\mathrm{Det}\ \overset{1}{A_1}) \ \ldots \ (\psi_w^{-1}\psi_{w^*})(\mathrm{Det}\ \overset{1}{A_{w^*}})\rho(\widetilde{1^*},\widetilde{\psi^*},U) = (\psi_w^{-1}\psi_1)(\mathrm{Det}\ \overset{1_*}{A_1^*}) \ \ldots \ (\psi_w^{-1}\psi_{w^*})(\mathrm{Det}\ \overset{1_*}{A_{w^*}^*}) \ . \tag{802}$$

Substitute (801), (802) into (800) and write A instead of A^*. Then

$$\Xi = (\mathrm{Det}\ Y)^{s_w+\frac{n^*}{4}} \sum_{A \in \mathcal{B}_\alpha(1;w,w^*)} (\psi_w^{-1}\psi_1)(\mathrm{Det}\ \overset{1}{A_1}) \ldots (\psi_w^{-1}\psi_{w^*})(\mathrm{Det}\ \overset{1}{A_{w^*}}) \sum_{U \in \Psi_\alpha(\widetilde{1^*})/\Delta_\alpha(\widetilde{1^*})} \times \tag{803}$$

$$\frac{1}{2} \int_{T \in \mathcal{F}_\alpha(\widetilde{1^*})} f(\widetilde{1^*},T[U],\widetilde{s^*}+(s_w+\tfrac{n}{4})\ e(w^*))S \exp(-\tfrac{\pi}{q}\ \mathrm{Tr}(Y[AW(n^*)]T[U])dv_T.$$

If V runs over $\Delta_\alpha(\widetilde{1^*})$ then

$$W(n^*)V'W(n^*) \tag{804}$$

runs over $\Delta_\alpha(1^*)$. Hence

$$\Xi = (\mathrm{Det}\ Y)^{s_w+\frac{n^*}{4}} \sum_{A \in \mathcal{B}_\alpha(1;w,w^*)/\Delta_\alpha(1^*)} (\psi_w^{-1}\psi_1)(\mathrm{Det}\ \overset{1}{A_1}) \ldots (\psi_w^{-1}\psi_{w^*})(\mathrm{Det}\ \overset{1}{A_{w^*}}) \times \tag{805}$$

$$\int_{T \in \mathcal{Y}(n^*)} f(\widetilde{1^*},T,\widetilde{s^*}+(s_w+\tfrac{n}{4})\ e(w^*))S \exp(-\tfrac{\pi}{q}\ \mathrm{Tr}(Y[AW(n^*)]T)dv_T \ .$$

Set

$$\text{(806)} \qquad \frac{\pi}{q}\, Y[AW(n^*)] = C^{-1}C^{-1'}$$

with an upper triangular matrix

$$\text{(807)} \qquad C = \begin{pmatrix} c_1 & & * \\ & \ddots & \\ 0 & & c_n \end{pmatrix} .$$

Then

$$\text{(808)} \qquad \widetilde{\frac{\pi}{q}\, Y[A]} = C'C .$$

From (432), (444), (808) we deduce

$$\text{(809)} \qquad f(\tilde{1}^*,T,s) = f(\tilde{1}^*,T[C^{-1}],s)f(\tilde{1}^*,\widetilde{\frac{\pi}{q}\, Y[A]},s) .$$

In the integral of (805) make the substitution $T \to T[C]$. Then

$$\text{(810)} \qquad \Xi = \Xi_1(\widetilde{s^*} + (s_w + \frac{n}{4})e(w^*))\Xi_2$$

with

$$\text{(811)} \qquad \Xi_1(s^*) = \int_{T \in \mathcal{Y}(n^*)} f(\tilde{1}^*,T,s^*)S \exp(-\mathrm{Tr}\ T)dv_T ,$$

$$\text{(812)} \qquad \Xi_2 = (\mathrm{Det}\ Y)^{s_w+\frac{n^*}{4}} \times$$

$$\sum_{A \in \mathcal{B}_\alpha(1;w,w^*)/\Delta_\alpha(1^*)} (\psi_w^{-1}\psi_1)(\mathrm{Det}\ \overset{1}{A}_1)\dots(\psi_w^{-1}\psi_{w^*})(\mathrm{Det}\ \overset{1}{A}_{w^*})f(\tilde{1}^*,\widetilde{\frac{\pi}{q}\, Y[A]},s^*-(s_w+\frac{n}{4})\, e(w^*)).$$

From (433) we deduce

$$\text{(813)} \qquad \Xi_2 = (\frac{\pi}{q})^{n^*(\{1^*,s^*\}-s_w-\frac{n}{4})}\ (\mathrm{Det}\ Y)^{s_w+\frac{n^*}{4}} \times$$

$$\sum_{A \in \mathcal{B}_\alpha(1;w,w^*)/\Delta_\alpha(1^*)} (\psi_w^{-1}\psi_1)(\mathrm{Det}\ \overset{1}{A}_1)\dots(\psi_w^{-1}\psi_{w^*})(\mathrm{Det}\ \overset{1}{A}_{w^*})f(1^*,Y[A],s^*-(s_w+\frac{n}{4})\, e(w^*)).$$

From theorem 29 and formulas (427), (428), (674), (766) we obtain

(814) $$\Xi_2 = \Xi_3\, \Lambda(l,\psi,Y,s)$$

(815) $$\Xi_3=\left(\frac{\Pi}{q}\right)^{n^*(\{l^*,s^*\}-s_w-\frac{n}{4})} \sum_{D \in \mathcal{J}^*(l;w,w^*)} (\psi_w^{-1}\psi_1)(\mathrm{Det}\ D_1)\ldots(\psi_w^{-1}\psi_{w^*})(\mathrm{Det}\ D_{w^*}) \times \prod_{\mu=1}^{w^*} (\mathrm{Det}\ D_\mu)^{2(s_\mu-s_w)+\frac{1}{2}(k_\mu+k_{\mu-1}-n^*-n)} .$$

Application of theorem 26 gives

(816) $$\Xi_3=\left(\frac{\Pi}{q}\right)^{n^*(\{l^*,s^*\}-s_w-\frac{n}{4})} \prod_{\nu=1}^{w^*} \prod_{\iota=0}^{l_\nu-1} L(\psi_w^{-1}\psi_\nu, 2(s_w-s_\nu)+\frac{l_w+l_\nu}{2}-\iota) .$$

In order to prove (783) we have to show

(817) $$\Xi_1(\widetilde{s^*} + (s_w + \tfrac{n}{4})e(w^*))\Xi_3 = \xi(l,\psi,s) .$$

Let $q > 1$. Then by theorem 40

(818) $$\Xi_1(s^*) = J(\widetilde{l^*},E,s^*)= \pi^{\frac{n^*(n^*-1)}{4}} \prod_{\nu=1}^{w^*} \prod_{\iota=0}^{\tilde{l}_\nu-1} \Gamma(s_\nu+\frac{\tilde{l}_\nu-n^*}{4}-\tfrac{\iota}{2}) ,$$

hence

(819) $$\Xi_1(\widetilde{s^*} + (s_w + \tfrac{n}{4})e(w^*)) = \pi^{\frac{n^*(n^*-1)}{4}} \prod_{\nu=1}^{w^*} \prod_{\iota=0}^{l_\nu-1} \Gamma(s_w-s_\nu+\frac{l_\nu+l_w}{4}-\tfrac{\iota}{2}).$$

From (816), (819) one gets (817) by an elementary computation.

Finally let $q = 1$. From (798), (811) and l^* instead of $\widetilde{l^*}$ we get

(820) $$\Xi_1(s^*) = \int_{T \in \mathcal{Y}(n^*)} f(l^*,T,s^*)P_{\frac{n}{2}}\exp(-\mathrm{Tr}\ T)\ dv_T .$$

Then (817) is equivalent to

(821) $$\Xi_1(s^*) = \pi^{\frac{n^*(n^*-1)}{4}} \prod_{\nu=1}^{w^*} \prod_{\iota=0}^{1_\nu-1} \Gamma(s_\nu+1+\frac{1_\nu-n^*}{4} - \frac{\iota}{2}) \times$$

$$\prod_{\nu=1}^{w^*} \prod_{\iota=0}^{1_\nu-1} (\frac{n}{2} - s_\nu + \frac{1_\nu-n^*}{4} - \frac{\iota}{2}) .$$

Hence it is left to prove (821).

We proceed like in Maaβ [33], page 83. From theorems 39, 40 we obtain for $X \in \mathcal{Y}(n)$

$$M_n(X)f(1,X,s) = M_n(\tilde{X}^{-1})f(\tilde{1},\tilde{X},\tilde{s}) =$$

$$\frac{1}{J(\tilde{1},E,\tilde{s})} \int_{Y \in \mathcal{Y}(n)} f(\tilde{1},Y,\tilde{s})M_n(\tilde{X}^{-1})\exp(-\mathrm{Tr}(\tilde{X}^{-1}Y))dv_Y =$$

$$\frac{(-1)^n(\mathrm{Det}\ X)\mathcal{J}(\tilde{1},E,\tilde{s}+e(w))}{J(\tilde{1},E,\tilde{s})} f(\tilde{1},\tilde{X},\widetilde{s-e(w)}) =$$

$$\frac{(-1)^n\mathcal{J}(\tilde{1},E,\tilde{s}+e(w))}{J(\tilde{1},E,\tilde{s})} f(1,X,s) .$$

Hence by (436)

(822) $$M_n(X)f(1,X,s)=(\prod_{\nu=1}^{w} \prod_{\iota=0}^{1_\nu-1}(s_\nu + \frac{n-1_\nu}{4} + \frac{\iota}{2}))f(1,X,s) .$$

From (400), (820), (822) we deduce

$$\Xi_1(s^*)= \int_{T \in \mathcal{Y}(n^*)} f(1^*,T,s^*)(\mathrm{Det}\ T)^{-\frac{n}{2}}\hat{M}_{n*}(\mathrm{Det}\ T)^{\frac{n}{2}+1}(\mathrm{Det}\frac{\partial}{\partial T}\exp(-\mathrm{Tr}\ T))dv_T =$$

$$(-1)^{n^*} \int_{T \in \mathcal{Y}(n^*)} (M_{n*}f(1^*,T,s^*-\frac{n}{2}e(w^*))(\mathrm{Det}\ T)^{\frac{n}{2}+1}\exp(-\mathrm{Tr}\ T)dv_T =$$

$$\left(\prod_{\nu=1}^{w}\prod_{\iota=0}^{l_\nu-1}\left(s_\nu - \frac{n}{2} + \frac{n^* - l_\nu}{4} + \frac{\iota}{2}\right)\right) J(l,E,s^*+e(w^*)) .$$

Now (821) follows from (436). Theorem 68 is proved.

Set

$$(823) \quad \check{\Xi}(l,\psi,s) = \prod_{1\leq\mu\leq\nu\leq w}\ \prod_{\iota=0}^{\max(l_\nu,l_\mu)-1} \xi\left(\psi_\nu^{-1}\psi_\mu, s_\nu - s_\mu + \frac{l_\nu+l_\mu}{4} - \frac{\iota}{2}\right) ,$$

$$(824) \quad \delta_1(l,s) = \prod_{1\leq\mu<\nu\leq w}\ \prod_{\iota=0}^{\max(l_\nu,l_\mu)-1} \left\{\left(s_\nu - s_\mu + \frac{l_\nu+l_\mu}{4} - \frac{\iota}{2}\right)\left(s_\mu - s_\nu + \frac{l_\nu+l_\mu}{4} - \frac{\iota}{2}\right)\right\} ,$$

$$(825) \qquad \Xi(l,\psi,s) = \begin{Bmatrix} \delta_1(l,s)\check{\Xi}(l,\psi,s) & (q = 1) \\ \check{\Xi}(l,\psi,s) & (q > 1) \end{Bmatrix} .$$

Then

$$(826) \qquad \check{\Xi}(\tilde{l},\tilde{\psi},\tilde{s}) = \check{\Xi}(l,\psi,s);\ \Xi(\tilde{l},\tilde{\psi},\tilde{s}) = \Xi(l,\psi,s) .$$

Set

$$(827) \qquad \check{\Phi}_o(l,\psi,s) = \check{\xi}(l,\psi,s)\check{\Xi}(l^*,\psi^*,s^*) ,$$

$$(828) \qquad \Phi_o(l,\psi,s) = \xi(l,\psi,s)\Xi(l^*,\psi^*,s^*) .$$

Then

$$(829) \qquad \Phi_o(l,\psi,s) = \delta(l,s)\delta_1(l^*,s^*)\check{\Phi}_o(l,\psi,s) \qquad (q = 1) ,$$

$$(830) \qquad \Phi_o(l,\psi,s) = \check{\Phi}_o(l,\psi,s) \qquad (q > 1) .$$

Put

$$(831)\quad \check{\Phi}(\mathfrak{l},\psi,s) = q^{-\frac{n^*(n^*-1)}{4}} \prod_{\nu=1}^{w^*} \prod_{\iota=\mathfrak{l}_\nu}^{\max(\mathfrak{l}_w,\mathfrak{l}_\nu)-1} \xi\left(\psi_w^{-1}\psi_\nu, s_w - s_\nu + \frac{\mathfrak{l}_w+\mathfrak{l}_\nu}{4} - \frac{\iota}{2}\right),$$

$$(832)\quad \delta_2(\mathfrak{l},s) = \prod_{\nu=1}^{w^*} \prod_{\iota=\mathfrak{l}_\nu}^{\max(\mathfrak{l}_w,\mathfrak{l}_\nu)-1} \left\{\left(s_w - s_\nu + \frac{\mathfrak{l}_w+\mathfrak{l}_\nu}{4} - \frac{\iota}{2}\right)\left(s_\nu - s_w + \frac{\mathfrak{l}_w+\mathfrak{l}_\nu}{4} - \frac{\iota}{2}\right)\right\}$$

$$(833)\quad \Phi(\mathfrak{l},\psi,s) = \begin{cases} \delta_2(\mathfrak{l},s)\check{\Phi}(\mathfrak{l},\psi,s) & (q = 1) \\ \check{\Phi}(\mathfrak{l},\psi,s) & (q > 1) \end{cases}.$$

Then

$$(834)\quad \check{\Xi}(\mathfrak{l},\psi,s) = \check{\Phi}_0(\mathfrak{l},\psi,s)\check{\Phi}(\mathfrak{l},\psi,s),$$

$$(835)\quad \Xi(\mathfrak{l},\psi,s) = \Phi_0(\mathfrak{l},\psi,s)\Phi(\mathfrak{l},\psi,s).$$

THEOREM 69: The function

$$(836)\quad \lambda(\mathfrak{l},\psi,Y,s) = \Phi_0(\mathfrak{l},\psi,s)(\mathrm{Det}\ Y)^{-\{\mathfrak{l},s\}}\Lambda(\mathfrak{l},\psi,Y,s)$$

is homogeneous in Y of degree 0. Furthermore

$$(837)\quad \Phi(\tilde{\mathfrak{l}},\tilde{\psi},\tilde{s})\lambda(\tilde{\mathfrak{l}},\tilde{\psi},\tilde{Y},\tilde{s}) = \Phi(\mathfrak{l},\psi,s)\lambda(\mathfrak{l},\psi,Y,s).$$

PROOF: Apply (765), (768), (826), (835).

THEOREM 70: Let $q > 1$ and ψ be a primitive character row, $\alpha = 1,2$ and

$$(838)\quad \lambda_{\alpha 1}(\mathfrak{l},\psi,Y,s) = \frac{1}{2}(\mathrm{Det}\ Y)^{s_2-\{\mathfrak{l},s\}} \int_{T\in\mathfrak{J}_{\alpha 1}(\widetilde{\mathfrak{l}^*})} \vartheta_\alpha(\mathfrak{l},\psi_2^{-1}\psi_1,Y,T)(\mathrm{Det}\ T)^{s_2-s_1}dv_T \qquad (w = 2),$$

$$(839)\qquad \lambda_{\alpha 1}(l,\psi,Y,s) = \frac{1}{2}\,\Phi(\widetilde{l^*},\widetilde{\psi^*},\widetilde{s^*})(\mathrm{Det}\ Y)^{s_w-\{l,s\}} \times$$

$$\int\limits_{T\in\mathcal{F}_{\alpha 1}(\widetilde{l^*})} \theta_\alpha(l,\psi_w^{-1}\psi^*,Y,T)(\mathrm{Det}\ T)^{s_w-\{l^*,s^*\}}\,\lambda(\widetilde{l^*},\widetilde{\psi^*},T,\widetilde{s^*})\,dv_T \qquad (w \geq 3).$$

The integrals on the right-hand-side of (838), (839) converge absolutely and are holomorphic for all $s \in \mathbb{C}^w$. Let R_n be a non-singular rational $n\times n$ matrix and

$$(840)\qquad Y = (y_{\iota\varkappa}) \in \mathcal{Y}(n,\mu)\ .$$

Let $\mathfrak{K} \subset \mathbb{C}^w$ be a compact domain. Then there exists a real number $c_{15} = c_{15}(l,\mu,R_n,\mathfrak{K}) > 1$ and finitely many (say $g(l)$) linear functions

$$(841)\qquad \mathcal{L}(l,\iota,\sigma) = \sum_{\nu=1}^{w} j(l,\iota,\nu)\sigma_\nu + j(l,\iota) \qquad (\iota = 1,\ldots,g(l))$$

with rational $j(l,\iota,\nu)$, $j(l,\iota)$ ($\iota = 1,\ldots,g(l)$; $\nu = 1,\ldots,w$), such that for $s \in \mathfrak{K}$ the inequality

$$(842)\qquad \mathrm{abs}\ \lambda_{\alpha 1}(l,\psi,Y[R_n],s) \leq c_{15}(\mathrm{Det}\ Y)^{\sigma_w-\{l,\sigma\}+\frac{n^*}{4}} \sum_{\iota=1}^{g(l)} y_1^{\mathcal{L}(l,\iota,\sigma)}$$

holds. By the formula

$$(843)\qquad \lambda(l,\psi,Y,s)= \lambda_{\alpha 1}(l,\psi,Y,s)+ \eta_\alpha(l,\psi)q^{2l_w(\{l,s\}-s_w)}\lambda_{\alpha 1}(\check{l},\check{\psi},\check{Y},\check{s})$$

the function $\lambda(l,\psi,Y,s)$ is holomorphically continued to $\mathbb{C}^w$. For $Y \in \mathcal{Y}(n,\mu)$ and $s \in \mathfrak{K}$ there is a real number $c_{16}= c_{16}(l,\mu,R_n,\mathfrak{K}) > 1$ with

$$(844)\qquad \mathrm{abs}\ \lambda(l,\psi,Y[R_n],s) \leq c_{15}(\mathrm{Det}\ Y)^{\sigma_w-\{l,\sigma\}+\frac{n^*}{4}} \sum_{\iota=1}^{g(l)} y_1^{\mathcal{L}(l,\iota,\sigma)}$$

$$+\ c_{16}(\mathrm{Det}\ Y)^{\sigma_w-\{l,\sigma\}-\frac{n^*}{4}} \sum_{\iota=1}^{g(l)} y_n^{-\mathcal{L}(\check{l},\iota,\check{\sigma})}\ .$$

Finally there is the functional equation

$$(845) \qquad \lambda(l,\psi,Y,s) = \eta_\alpha(l,\psi) q^{2l_w(|l,s|-s_w)} \check{\lambda}(\check{l},\check{\psi},\check{Y},\check{s}) .$$

Hence

$$(846) \qquad \eta_1(l,\psi) = \eta_2(l,\psi) .$$

For $q = 1$ the same is true for $w = 2$ or $w \geq 3$ and

$$(847) \qquad l_1 = \dots = l_w .$$

If (847) does not hold, we can only prove that $\lambda_{\alpha 1}(l,\psi,Y,s)$ and $\lambda(l,\psi,Y,s)$ are meromorphic in $\mathbb{C}^w$. They become holomorphic if one multiplies them by a finite product of linear functions in s.

PROOF: Because of (840) there is a constant $d_1 \geq 1$ with

$$(848) \qquad d_1^{-1}(\mathrm{Dg}\ Y) \leq Y \leq d_1(\mathrm{Dg}\ Y).$$

Hence there exists a constant $d_2 = d_2(n,\mu) \geq d_1 \geq 1$ with

$$(849) \qquad j(Y) = d_2^{-1} y_1,\ j(\check{Y}) = d_2^{-1} y_n^{-1} .$$

Finally we have

$$(850) \qquad (Y[R_n])^\vee = \check{Y}[\check{R}_n]$$

with the non-singular rational nxn matrix $\check{R}_n = QR_n'^{-1}Q^{-1}$.

Let

$$(851) \qquad T = (t_{\iota\varkappa}) \in \mathcal{M}(n^*) .$$

Then by Minkowski's reduction theory there is a constant $d_3 > 1$ with

(852) $$d_3^{-1} t_1 \leq t_\iota \leq d_3 t_{n^*} \qquad (\iota = 1,\ldots,n^*),$$

(853) $$\operatorname{abs} t_{\iota\varkappa} \leq d_3 t_{n^*} \qquad (\iota \neq \varkappa;\ \iota,\varkappa = 1,\ldots,n^*),$$

(854) $$d_3^{-1}\, t_1^{n^*-1} t_{n^*} \leq \operatorname{Det} T \leq d_3 t_1 t_{n^*}^{n^*-1} \quad ,$$

(855) $$(\operatorname{Det} T)^{\frac{1}{n^*}} \leq d_3 t_{n^*} \ .$$

Furthermore

(856) $$\operatorname{Tr} T \geq t_{n^*} \quad .$$

Now let $T \in \mathfrak{M}_1(n^*)$, i. e.,

(857) $$\operatorname{Det} T \geq 1 \ .$$

Then by (854), (855) we obtain

(858) $$t_1 \geq d_3^{-1}\ t_{n^*}^{1-n^*} \quad ,$$

(859) $$t_{n^*} \geq d_3^{-1} \ .$$

From (852), (854), (858) we deduce

(860) $$d_3^{-2}\ t_{n^*}^{1-n^*} \leq t_\iota \leq d_3 t_{n^*} \qquad (\iota = 1,\ldots,n^*) \ ,$$

(861) $$d_3^{-n^*} t_{n^*}^{1-(n^*-1)^2} \leq \operatorname{Det} T \leq d_3^{n^*}\ t_{n^*}^{n^*} \quad .$$

Hence for all real numbers p

(862) $$t_\iota^p \leq d_3^{2 \operatorname{abs} p}(t_{n^*}^p + t_{n^*}^{(1-n^*)p}) \qquad (\iota = 1,\ldots,n^*),$$

(863) $$(\operatorname{Det} T)^p \leq d_3^{n^* \operatorname{abs} p}(t_{n^*}^{n^* p} + t_{n^*}^{(1-(n^*-1)^2)p}) \ .$$

Now let $w = 2$. From (373), (838) we get with some constant $d_4 > 1$:

$$(864)\qquad \operatorname{abs} \lambda_{\alpha 1}(1,\psi,Y[R_n],s) \le d_4(\operatorname{Det} Y)^{\sigma_2-\{1,\sigma\}} \times$$

$$\sum_{\nu=1}^{h_\alpha} \int_{T \in \mathcal{M}_1(n^*)} \operatorname{abs} \theta_\alpha(1,\psi_2^{-1}\psi_1,Y[R_n],T[F_{\alpha\nu}])(\operatorname{Det} T)^{\sigma_2-\sigma_1}\, dv_T \quad .$$

From theorem 53 and formulas (849), (856), (863) we get

$$(865)\qquad \operatorname{abs} \lambda_{\alpha 1}(1,\psi,Y[R_n],s) \le$$

$$d_5(\operatorname{Det} Y)^{\sigma_2-\{1,\sigma\}+\frac{n^*}{4}}\, y_1^{-\frac{nn^*}{2}} \sum_{\iota} \int_{T \in \mathcal{M}_1(n^*)} t_{n^*}^{\beta_1(\iota,\sigma)} \exp(-d_6^{-1} y_1 t_{n^*})[dT]$$

with finitely many linear functions $\beta_1(\iota,\sigma)$. According to (853), (859), (860) we integrate over all $t_{\iota\varkappa}$ except t_{n^*}. This gives

$$(866)\qquad \operatorname{abs} \lambda_{\alpha 1}(1,\psi,Y[R_n],s) \le d_7(\operatorname{Det} Y)^{\sigma_2-\{1,\sigma\}+\frac{n^*}{4}} \times$$

$$y_1^{-\frac{nn^*}{2}} \sum_{\iota} \int_{d_3^{-1}}^{\infty} t_{n^*}^{\beta_2(\iota,\sigma)} \exp(-d_6^{-1}\, y_1\, t_{n^*})\, dt_{n^*}$$

with finitely many linear functions $\beta_2(\iota,\sigma)$. Like in (112) set for $m > 0$; $p \in \mathbb{R}$:

$$(867)\qquad I(m,p) = \int_m^\infty u^p \exp(-u)\frac{du}{u} \; .$$

Then theorem 12 gives for $\epsilon > 0$

$$(868)\qquad I(m,p) \le c_5(m^0 + m^{p-\epsilon} + m^{p+\epsilon}) \quad .$$

Substitute in the integrals of (866) the variable $u = d_6^{-1} y_1 t_{n^*}$. Then each integral becomes of type (867). Applying (868) we get (842). Hence $\lambda_{\alpha 1}(1,\psi,Y,s)$ is holomorphic in $\mathfrak{K}$. But if $s \in \mathbb{C}^w$ and $Y \in \mathcal{Y}(n)$ there are $\mathfrak{K}$ and μ with $s \in \mathfrak{K}$, $Y \in \mathcal{T}(n,\mu)$.

Hence for each $Y \in \mathfrak{Y}(n)$ the function $\lambda_{\alpha 1}(1,\psi,Y,s)$ is integral in s.

From (828) we obtain $\Phi_0(1,\psi,s) = \xi(1,\psi,s)$ for $w = 2$. Hence by (780), (836)

$$(869)\quad \lambda(1,\psi,Y,s) = \frac{1}{2}(\operatorname{Det} Y)^{s_2-\{1,s\}} \int_{T \in \widetilde{\mathfrak{f}_\alpha(1^*)}} \theta_\alpha(1,\psi_2^{-1}\psi_1,Y,T)(\operatorname{Det} T)^{s_2-s_1} dv_T \qquad (w = 2).$$

Because of (390), (413), (414), (423), (838), (869) therefore

$$(870)\qquad \lambda(1,\psi,Y,s) = \lambda_{\alpha 1}(1,\psi,Y,s) +$$

$$q^{2l_w(\{1,s\}-s_2)}(\operatorname{Det} \check{Y})^{\check{s}_2-\{\check{1},\check{s}_2\}} \int_{\substack{T \in \widetilde{\mathfrak{f}_\alpha(1^*)} \\ \operatorname{Det} T \leq 1}} \theta_\alpha(1,\psi_2^{-1}\psi_1,Y,T)(\operatorname{Det} T)^{\check{s}_1-\check{s}_2} dv_T \qquad (w = 2).$$

Let $T \in \mathfrak{Y}(n^*)$ and $U \in \Psi_\alpha(1^*)$. Then $\widetilde{T[U]} = \check{T}[\check{U}]$. Applying theorem 30 we see that the integration over $T \in \{\widetilde{\mathfrak{f}_\alpha(1^*)};\ \operatorname{Det} T \leq 1\}$ is the same as an integration over $\check{T} \in \mathfrak{f}_{\alpha 1}(1^*)$. Hence from theorem 47 and (870) we obtain (843). From (842), (843) we deduce (844), (845), (846). For $w = 2$ the theorem is proved.

Let $w \geq 3$ and assume that the theorem is true for $w-1 = w^*$.

Let $q > 1$. By theorem 66 the function $\xi(\chi,s)$ is holomorphic. Hence by (831) and (833) the function $\Phi(1,\psi,s)$ is holomorphic. Let $q = 1$ and (847) hold. Then by (831) we have $\check{\Phi}(1,\psi,s) = 1$ and by (832), (833) the function $\Phi(1,\psi,s)$ is again holomorphic.

Apply theorem 53 to (839) and estimate $\lambda(\widetilde{1^*},\widetilde{\psi^*},T,\widetilde{s^*})$ by (844) with w^* instead of w. Since $\Phi(\widetilde{1^*},\widetilde{\psi^*},\widetilde{s^*})$ is holomorphic it is bounded in $\mathfrak{k}$. Now we get (842) like in the case $w = 2$. From theorem 47 and formula (837) we get (843) and hence (844), (845), (846).

Now let $q = 1$ and suppose that (847) does not hold. By theorem 66 the product $s\xi(\chi,s)$ has a pole of first order at $s = \frac{1}{2}$. Because of (831), (832), (833) the function $\Phi(1,\psi,s)$ is not holomorphic but it may be made holomorphic by multiplying it with certain finitely many linear functions in s. Hence one can prove with the former method that $\lambda_{\alpha 1}(1,\psi,Y,s)$ and $\lambda(1,\psi,Y,s)$ are meromorphic in $\mathbb{C}^w$ and become holomorphic by multiplication with certain finitely many linear functions in s. Theorem 70 is proved.

THEOREM 71: Let $m \in \mathbb{N}$ and χ be an even primitive character mod q. Then

$$G(m,\chi) = (G(1,\chi))^m \quad . \tag{871}$$

PROOF: Apply (500), (501), (846).

In [33], page 220 upper part Maaβ considers a homogeneous polynomial $w_o(x)$ of degree gn with even g. Now let $g \in \mathbb{Z}$ be arbitrary and χ a character mod q with

$$\chi(-1) = (-1)^g \ . \tag{872}$$

Instead of Maaβ's function $\Phi_o(\ldots)$ (page 220) consider the sum

$$\sum_{A = (\overset{1}{A}_{\nu\mu}) \in \mathcal{B}_\alpha(1;2,1)/\Delta_\alpha(1^*)} \chi(\operatorname{Det} A_1) \frac{w_o(CA)}{(\operatorname{Det} Y[A])^{s+\frac{g}{2}}} \quad (\alpha=1,2) \tag{873}$$

with $Y = C'C \in \mathfrak{Y}(n)$. From theorem 29 it follows that (873) does not depend on α. I conjecture the following. With a mixture of the methods of Maaβ [33] and this manuscript one sees that (873) has analytic continuation and satisfies a functional equation. We get an equation of type (871) for all primitive characters χ χ mod q (also odd ones).

THEOREM 72: Let $q > 1$ and

(874) $$\mathfrak{m}(l) = \min(l_1,\ldots,l_w) \geq 1.$$

In the domain

(875) $$\sigma_{\nu+1} - \sigma_\nu \geq \frac{l_{\nu+1}+ l_\nu}{4} - \frac{1}{2}\mathfrak{m}(l) \qquad (\nu = 1,\ldots,w-1)$$

the functions $\xi(l,\psi,s)$, $\Xi(l,\psi,s)$, $\Phi_o(l,\psi,s)$, $\Phi(l,\psi,s)$ have no zeros.

PROOF: Apply theorem 66.

THEOREM 73: Let $q > 1$. Then the functions $\Lambda(l,\psi,Y,s)$, $\zeta(l,\psi,Y,s)$, $\zeta^*(l,\psi,Y,s)$ are meromorphic in $\mathbb{C}^w$. They are holomorphic in the domain (875).

PROOF: Use theorems 70, 72.

THEOREM 74: Let $q > 1$ and

(876) $$l_1 = l_2 = \ldots = l_w .$$

Then $\Lambda(l,\psi,Y,s)$ and $\zeta^*(l,\psi,Y,s)$ are holomorphic in the domain

(877) $$\sigma_{\nu+1} - \sigma_\nu \geq 0 \qquad (\nu = 1,\ldots,w-1).$$

$\zeta(l,\psi,Y,s)$ is holomorphic in $\mathbb{C}^w$. Furthermore

(878) $$\Phi(l,\psi,s) = q^{-\frac{n^*(n^*-1)}{4}} .$$

PROOF: The first part follows from theorem 73 and (878) from (831), (833). From (769), (836) we deduce

(879) $$\zeta(l,\psi,Y,s) = \frac{L(l,\psi,s)}{\Phi_o(l,\psi,s)}(\operatorname{Det} Y)^{\{l,s\}-s_w-\frac{n^*}{4}}\,\lambda(l,\psi,Y,s).$$

In the case (876)

(880) $$\Phi_o(l,\psi,s) = q^{\frac{n^*(n^*-1)}{4}} \prod_{1\le\mu<\nu\le w}\ \prod_{\iota=0}^{l_1-1} \xi(\psi_\nu^{-1}\psi_\mu, s_\nu - s_\mu + \frac{l_1-\iota}{2}),$$

(881) $$L(l,\psi,s) = \prod_{1\le\mu<\nu\le w}\ \prod_{\iota=0}^{l_1-1} L(\psi_\nu^{-1}\psi_\mu, 2(s_\nu-s_\mu) + l_1-\iota).$$

From (770), (880), (881) we see that $\frac{L(l,\psi,s)}{\Phi_o(l,\psi,s)}$ is holomorphic for $s \in \mathbb{C}^w$. Because of theorem 70 and (879) the function $\zeta(l,\psi,Y,s)$ is holomorphic for $s \in \mathbb{C}^w$. Theorem 74 is proved.

THEOREM 75: Let $q > 1$. The functions $\hat{\zeta}(l,\chi,Y,z)$, $\hat{\zeta}^*(l,\chi,Y,z)$ are meromorphic for $z \in \mathbb{C}^{w-1}$. They are holomorphic in the domain

(882) $$\operatorname{Re} z_\nu \ge \frac{l_{\nu+1}+l_\nu}{2} - \frac{1}{2}\,\mathfrak{m}(l) \qquad (\nu = 1,\ldots,w-1).$$

PROOF: Apply theorem 73.

THEOREM 76: Let $q > 1$ and (876) be true. Then $\hat{\zeta}^*(l,\chi,Y,z)$ is holomorphic in the domain

(883) $$\operatorname{Re} z_\nu \ge \frac{l_1}{2} \qquad (\nu = 1,\ldots,w-1).$$

$\hat{\zeta}(l,\chi,Y,z)$ is holomorphic in $\mathbb{C}^{w-1}$.

PROOF: Apply theorem 74.

§ 12. FUNCTIONAL EQUATIONS

For Selberg's zeta- and L-series functional-equations are proved like in Maaß [33], § 17 and Terras [45], [46]. For Selberg's L-series the same types of functional equations are true as for Selberg's zetafunctions, but for L-series the expressions become more complicated.

From (845) we deduce

(884) $$\lambda(l,\psi,Y,s) = \eta_1(l,s)q^{2l_w(\{l,s\}-s_w)}\lambda(\check{l},\check{\psi},\check{Y},\check{s}) .$$

From (837) we get

(885) $$\lambda(l,\psi,Y,s) = \frac{\Phi(\tilde{l},\tilde{\psi},\tilde{s})}{\Phi(l,\psi,s)}\lambda(\tilde{l},\tilde{\psi},\tilde{Y},\tilde{s}) .$$

Set

(886) $$\mathfrak{z}Y = Y[P(l)^{-1}] .$$

Then

(887) $$\mathfrak{z}l = \tilde{\check{l}},\ \mathfrak{z}\psi = \tilde{\check{\psi}},\ \mathfrak{z}s = \tilde{\check{s}},\ \mathfrak{z}Y = q^{-\frac{2l_w}{n}}\tilde{\check{Y}} .$$

Since $\lambda(l,\psi,Y,s)$ is homogeneous in Y of degree 0 we get

(888) $$\lambda(l,\psi,Y,s) = \gamma_1(l,\psi,s)\lambda(\mathfrak{z}l,\mathfrak{z}\psi,\mathfrak{z}Y,\mathfrak{z}s)$$

with

(889) $$\gamma_1(l,\psi,s) = \frac{\Phi(\tilde{\check{l}},\tilde{\check{\psi}},\tilde{\check{s}})}{\Phi(\check{l},\check{\psi},\check{s})}\eta_1(l,\psi)q^{2l_w(\{l,s\}-s_w)} .$$

THEOREM 77: Let $w \geq 3$ and set

$$(890)\qquad \gamma_2(l,\psi,s) = \frac{\Phi(\widetilde{l^*},\widetilde{\psi^*},\widetilde{s^*})}{\Phi(\widetilde{\mathfrak{z}^*l^*},\widetilde{\mathfrak{z}^*\psi^*},\widetilde{\mathfrak{z}^*s^*})}\,\delta_1^{-1}(\widetilde{\mathfrak{z}^*l^*},\widetilde{\mathfrak{z}^*\psi^*},\widetilde{\mathfrak{z}^*s^*}) \times q^{-\frac{2l_w k_{w-2}}{n^*}(s_w - \{l,s\} + \frac{n^*}{4})} .$$

Then

$$(891)\qquad \lambda_{11}(l,\psi,Y,s) = \gamma_2(l,\psi,s)\sum_{\rho=1}^{r}\lambda_{11}(\hat{l},\hat{\psi},Y[K_\rho],\hat{s}) ,$$

$$(892)\qquad \lambda_{11}(\check{l},\check{\psi},\check{Y},\check{s}) = \gamma_2(l,\psi,s)\sum_{\rho=1}^{r}\lambda_{11}(\check{\hat{l}},\check{\hat{\psi}},(Y[K_\rho])^{\vee},\check{\hat{s}}) ,$$

$$(893)\qquad \lambda(l,\psi,Y,s) = \gamma_2(l,\psi,s)\sum_{\rho=1}^{r}\lambda(\hat{l},\hat{\psi},Y[K_\rho],\hat{s}) .$$

PROOF: Because of $\sim\, = \mathfrak{z}^{\sim}\mathfrak{z}$ and $\mathfrak{z}^{-1}Y = Y[P(l)]$, we get

$$(894)\qquad \lambda(\widetilde{l^*},\widetilde{\psi^*},T,\widetilde{s^*}) = \lambda(\mathfrak{z}^*(\widetilde{\mathfrak{z}^*l^*}),\mathfrak{z}^*(\widetilde{\mathfrak{z}^*\psi^*}),\mathfrak{z}^*T[P(\widetilde{\mathfrak{z}^*l^*})],\mathfrak{z}^*(\widetilde{\mathfrak{z}^*s^*})).$$

From (887), (894) we deduce

$$(895)\qquad \lambda(\widetilde{l^*},\widetilde{\psi^*},T,\widetilde{s^*}) = \delta_1^{-1}(\widetilde{\mathfrak{z}^*l^*},\widetilde{\mathfrak{z}^*\psi^*},\widetilde{\mathfrak{z}^*s^*})\lambda(\widetilde{\mathfrak{z}^*l^*},\widetilde{\mathfrak{z}^*\psi^*},T[P(\widetilde{\mathfrak{z}^*l^*})],\widetilde{\mathfrak{z}^*s^*}).$$

From (223), (242), (328) we deduce

$$(896)\qquad P(\widetilde{\mathfrak{z}^*l^*}) = P(l^*) .$$

From (333) we get

$$(897)\qquad P(\widetilde{\mathfrak{z}^*l^*})\Psi_1(\widetilde{\mathfrak{z}^*l^*})P^{-1}(\widetilde{\mathfrak{z}^*l^*}) = \Psi_1(\widetilde{l^*}) .$$

Hence if T runs over a fundamental domain of $\Psi_1(\widetilde{l^*})$ the matrix

(898) $$T^* = T[P(\widetilde{\mathfrak{z}*1^*})] = T[P(1^*)]$$

runs over a fundamental domain of $\Psi_1(\widetilde{\mathfrak{z}*1^*})$.

Now insert (895) into (839), make the variable transformation $T \to T^*$ and apply (576). Then (891) follows. In the same way (892) follows from (577), Using $\eta(\hat{1},\hat{\psi}) = \eta(1,\psi)$ formula (893) follows from (843), (891), (892). Theorem 77 is proved.

Obviously the permutations $\mathfrak{z}$ and $\wedge$ generate the symmetric group Υ_W. So from (888), (893) one may deduce the behaviour of $\lambda(1,\psi,Y,s)$ under all permutations.

§ 13. RESIDUES OF SELBERG'S ZETAFUNCTIONS

Riemann's zetafunction has a pole at $s = 1$. Dirichlet's L-series are holomorphic everywhere. A similar situation is true for Selberg's zeta- and L-series. In § 11 we saw that Selberg's L-series have nice holomorphy properties. Selberg's zetafunction has more poles and one may compute residues. Details about this residue computation are given in Maaß [33], § 17.

From now on we consider only the case $q = 1$. It will turn out that then the functions $\Lambda(1,\psi,Y,s)$, $\zeta(1,\psi,Y,s)$, $\zeta^*(1,\psi,Y,s)$ are not holomorphic in (875). For $q = 1$ the only character is the principal character. Hence we set

(899) $$\zeta(1,Y,s) = \zeta(1,\psi,Y,s);\ \zeta^*(1,Y,s) = \zeta^*(1,\psi,Y,s) ;$$

(900) $$\Lambda(1,Y,s) = \Lambda(1,\psi,Y,s);\ \lambda(1,Y,s) = \lambda(1,\psi,Y,s) ;$$

(901) $$L(1,s) = L(1,\psi,s);\ \Xi(1,s) = \Xi(1,\psi,s) ;$$

(902) $$\Phi_0(1,s) = \Phi_0(1,\psi,s);\ \Phi(1,s) = \Phi(1,\psi,s) ;$$

(903) $$\gamma_\nu(1,s) = \gamma_\nu(1,\psi,s)\ (\nu = 1,2);\ \xi(1,s) = \xi(1,\psi,s);$$

(904) $$\hat{\zeta}(1,Y,u) = \hat{\zeta}(1,\chi,Y,u);\ \hat{\zeta}^*(1,Y,u) = \hat{\zeta}^*(1,\chi,Y,u)\ ;$$

(905) $$\hat{L}(1,u) = \hat{L}(1,\chi,u)\ .$$

Then

(906) $$\hat{\zeta}(1,Y,u) = \sum_{\{\alpha\}} \prod_{\beta=1}^{w-1} (\mathrm{Det}(Y[B_{w-1,\beta}]))^{-u_\beta}\ ;$$

(907) $$\hat{\zeta}^*(1,Y,u) = \sum_{U \in \Omega(n)/\Delta(1)} \prod_{\beta=1}^{w-1} (\mathrm{Det}((Y[U])_{k_\beta}))^{-u_\beta}$$

with $\Delta(1) = \Delta_\alpha(1)$ $(\alpha = 1,2)$;

(908) $$\hat{L}(1,u) = \prod_{1 \leq \mu \leq \nu \leq w-1} \prod_{\varkappa=0}^{l_\mu - 1} \zeta(2(u_\mu + \ldots + u_\nu) - k_\nu + k_\mu - \varkappa)\ ,$$

where ζ is Riemann's zetafunction;

(909) $$\hat{\zeta}(1,Y,u) = \hat{L}(1,u)\hat{\zeta}^*(1,Y,u)\ .$$

THEOREM 78: The functions (906), (907) are homogeneous in Y of degree

(910) $$-\sum_{\mu=1}^{w-1} k_\mu u_\mu\ .$$

They are invariant under

(911) $$Y \to Y[U] \qquad (U \in \Omega(n))\ .$$

PROOF: Apply theorem 63.

We have

(912) $$\zeta(1,Y,s) = L(1,s)\zeta^*(1,Y,s)\ ;$$

$$(913)\qquad \zeta(l,Y,s) = \sum_{\{\mathfrak{a}\}} \left(\prod_{\nu=1}^{w-1} (\mathrm{Det}(Y[B_{w-1,\nu}]))^{s_\nu - s_{\nu+1} - \frac{l_{\nu+1}+l_\nu}{4}}\right) ;$$

$$(914)\qquad \zeta^*(l,Y,s) = \sum_{U \in \Omega(n)/\Delta(l)} \left(\prod_{\nu=1}^{w-1} (\mathrm{Det}((Y[U])_{k_\nu}))^{s_\nu - s_{\nu+1} - \frac{l_{\nu+1}+l_\nu}{4}}\right) ;$$

$$(915)\qquad L(l,s) = \prod_{1 \le \mu \le \nu \le w-1} \prod_{\varkappa=0}^{l_\mu - 1} \zeta(2(s_{\nu+1}-s_\mu) + \tfrac{1}{2}(l_{\nu+1}+l_\mu) - \varkappa)$$

where again ζ is Riemann's zetafunction.

THEOREM 79: The functions (913), (914) are homogeneous in Y of degree

$$(916)\qquad n(\{l,s\} - s_w - \tfrac{1}{4} k_{w-1}) .$$

They are invariant under the substitution (911).

PROOF: Apply theorem 64.

THEOREM 80: The function

$$(917)\qquad \Lambda(l,Y,s) = (\mathrm{Det}\ Y)^{s_w + \frac{n^*}{4}} \zeta^*(l,Y,s)$$

is homogeneous in Y of degree $n\{l,s\}$ and invariant under (911).

PROOF: Apply theorem 65.

One has

$$(918)\qquad \Lambda(l,Y,s) = \sum_{U \in \Omega(n)/\Delta(l)} f(l,Y[U],s);$$

$$(919)\qquad \Lambda(\tilde{l},\tilde{Y},\tilde{s}) = \Lambda(l,Y,s) ;$$

(920) $(\operatorname{Det} Y)^{s_w + \frac{n^*}{4}} \zeta(1,Y,s) = L(1,s)\Lambda(1,Y,s)$.

THEOREM 81: Let $H(1,Y,\ldots)$ be one of the functions (899), (900), (904). Then

(921) $H(1,Y^{-1},\ldots) = H(1,\tilde{Y},\ldots) = H(1,\check{Y},\ldots)$.

PROOF: For $q = 1$ we have

(922) $W(n),Q(1),P(1) \in \Omega(n)$.

Now (921) follows from

(923) $H(1,Y[U],\ldots) = H(1,Y,\ldots) \quad (U \in \Omega(n))$.

Instead of (770) we get

(924) $\xi(s) = \pi^{-s}\,\Gamma(s)\zeta(2s)$,

where ζ is Riemann's zetafunction.

THEOREM 82: The function

(925) $s(\frac{1}{2} - s)\xi(s)$

is holomorphic in $\mathbb{C}$ and invariant under $s \to \frac{1}{2} - s$. All zeros lie in the strip $0 < \operatorname{Re} s < \frac{1}{2}$.

PROOF: See theorem 66 and Landau [22], § 128.

The function

(926) $\mathring{\xi}(s) = s\xi(s)$

is holomorphic in $\mathbb{C}$ except a pole of first order at $s = \frac{1}{2}$. From (776), (777), (778) we deduce

$$(927)\qquad \xi(l,s)=\prod_{\nu=1}^{w^*}\prod_{\iota=0}^{l_\nu-1}\{\mathring{\xi}(s_w - s_\nu + \frac{l_w+l_\nu}{4} - \frac{\iota}{2})(s_\nu - s_w + \frac{l_w+l_\nu}{4} - \frac{\iota}{2})\}.$$

$$(928)\qquad \Xi(l,s) = \prod_{1\le\mu<\nu\le w}\prod_{\iota=0}^{\max(l_\nu,l_\mu)-1}\{\mathring{\xi}(s_\nu - s_\mu + \frac{l_\nu+l_\mu}{4} - \frac{\iota}{2})(s_\mu - s_\nu + \frac{l_\nu+l_\mu}{4} - \frac{\iota}{2})\}.$$

Then

$$(929)\qquad \Xi(\tilde{l},\tilde{s}) = \Xi(l,s) ;$$

$$(930)\qquad \Phi_0(l,s) = \xi(l,s)\Xi(l^*,s^*) ;$$

$$(931)\qquad \Phi(l,s)=\prod_{\nu=1}^{w^*}\prod_{\iota=l_\nu}^{\max(l_w,l_\nu)-1}\{\mathring{\xi}(s_w - s_\nu + \frac{l_w+l_\nu}{4} - \frac{\iota}{2})(s_\nu - s_w + \frac{l_w+l_\nu}{4} - \frac{\iota}{2})\},$$

$$(932)\qquad \Xi(l,s) = \Phi_0(l,s)\Phi(l,s) .$$

THEOREM 83: The function

$$(933)\qquad \lambda(l,Y,s) = \Phi_0(l,s)\cdot(\mathrm{Det}\ Y)^{-\{l,s\}}\Lambda(l,Y,s)$$

is homogeneous in Y of degree 0. Furthermore

$$(934)\qquad \Phi(\tilde{l},\tilde{s})\lambda(\tilde{l},\tilde{Y},\tilde{s}) = \Phi(l,s)\lambda(l,Y,s) .$$

$$(935)\qquad \lambda(l,Y,s) = \lambda(\check{l},\check{Y},\check{s}) .$$

PROOF: Apply theorems 69, 70.

From (886), (888), (889), (890), (922), (923) we get

$$(936)\qquad \lambda(l,Y,s) = \gamma_1(l,s)\lambda(\mathfrak{z}l,Y,\mathfrak{z}s) ,$$

$$(937) \qquad \gamma_1(l,s) = \frac{\Phi(\tilde{\check{l}}, \tilde{\check{s}})}{\Phi(\check{l}, \check{s})} ,$$

$$(938) \qquad \gamma_2(l,s) = \frac{\Phi(\tilde{l}^*, \tilde{s}^*)\,\Phi(\check{\widetilde{\mathfrak{z}^* l^*}}, \check{\widetilde{\mathfrak{z}^* s^*}})}{\Phi(\widetilde{\mathfrak{z}^* l^*}, \widetilde{\mathfrak{z}^* s^*})\,\Phi(\widetilde{\check{\mathfrak{z}^* l^*}}, \widetilde{\check{\mathfrak{z}^* s^*}})} .$$

Applying (887) we get

$$(939) \qquad \gamma_2(l,s) = \frac{\Phi(l^*, s^*)}{\Phi(\check{l}^*, \check{s}^*)} .$$

For $q = 1$ we have $r = 1$ and $K_1 \in \Omega(n)$. Hence (893) gives

$$(940) \qquad \lambda(l,Y,s)\gamma_2(l,s)\lambda(\hat{l},Y,\hat{s}) .$$

THEOREM 84: For a given integer ρ in $1 \leq \rho \leq w^* = w-1$ the function

$$(941) \qquad (u_\rho - \tfrac{1}{2}(l_{\rho+1} + l_\rho))\zeta^*(l,Y,s)$$

is holomorphic in the domain defined by

$$(942) \qquad \operatorname{abs}(u_\rho - \tfrac{1}{2}(l_{\rho+1} + l_\rho)) < \tfrac{1}{4}, \quad \operatorname{Re} u_\nu > n \ (1 \leq \nu \leq w^*,\ \nu \neq \rho).$$

Its value at $u_\rho = \frac{1}{2}(l_{\rho+1} + l_\rho)$ is, up to a positive constant factor, equal to

$$(943) \qquad \left\{ \begin{array}{ll} \zeta^*(\hat{l},Y,\hat{s}) & (\rho < w^*) \\ \zeta^*(\hat{l},Y,\hat{s})(\operatorname{Det} Y)^{-\frac{1}{2}l_{w^*}} & (\rho = w^*) \end{array} \right\}$$

and

$$(944) \qquad \hat{l} = (\hat{l}_1, \ldots, \hat{l}_{w^*});\ \hat{s} = (\hat{s}_1, \ldots, \hat{s}_{w^*}) ,$$

(945) $$\hat{l}_\nu = \left\{ \begin{array}{ll} l_\nu & (1 \leq \nu < \rho) \\ l_\rho + l_{\rho+1} & (\nu = \rho) \\ l_{\nu+1} & (\rho < \nu \leq w^*) \end{array} \right\} ,$$

(946) $$\hat{s}_\nu = \left\{ \begin{array}{ll} s_\nu & (1 \leq \nu < \rho) \\ s_\rho + \frac{1}{4} l_{\rho+1} & (\nu = \rho) \\ s_{\nu+1} & (\rho < \nu \leq w^*) \end{array} \right\} .$$

PROOF: See Maaβ [33], page 287, theorem 3 or Terras [45], [46].

From (917), (933) we deduce

(947) $$\lambda(l,Y,s) = \Phi_o(l,s)(\operatorname{Det} Y)^{s_w - \{l.s\} + \frac{n^*}{4}} \zeta^*(l,Y,s) .$$

An easy computation shows

(948) $$\{l,s\} = \{\hat{l},\hat{s}\} + \frac{1}{4} l_\rho l_{\rho+1} .$$

Then theorem 84 gives

(949) $$\lim_{s_{\rho+1} \to s_\rho + \frac{1}{4}(l_\rho + l_{\rho+1})} ((\operatorname{Det} Y)^{s_w - \{l,s\} + \frac{n^*}{4}} \zeta^*(l,Y,s)) =$$

$$(\operatorname{Det} Y)^{\hat{s}_{w^*} - \{\hat{l},\hat{s}\} + \frac{k_{w^*-1}}{4} - \frac{l_\rho l_{\rho+1}}{4}} \zeta^*(\hat{l},Y,\hat{s}) \qquad (\rho = 1,\dots,w^*).$$

THEOREM 85: Let $\mathring{l}, p$ be defined by (222), (430), (431). Set

(950) $$\beta(l,s) =$$

$$\prod_{1 \leq \mu < \nu \leq w} \prod_{\alpha=o}^{l_\mu - 1} \prod_{\beta=o}^{l_\nu - 1} \{\mathring{\xi}(s_\nu - s_\mu + \frac{l_\mu + l_\nu}{4} - \frac{(\alpha+\beta)}{2})(s_\mu - s_\nu + \frac{l_\mu + l_\nu}{4} - \frac{(\alpha+\beta)}{2})\} .$$

Then

$$(951)\qquad \lambda(\overset{\circ}{l},Y,p) = \text{const}\cdot\beta(l,s)(\text{Det } Y)^{-\{l,s\}}\Lambda(l,Y,s)\ .$$

PROOF: Apply theorem 84.

By theorem 70 the left-hand-side of (951) is an integral function in s. The factors of $\Phi_0(l,s)$ are contained among the factors of $\beta(l,s)$. Hence by (933) there is an $\alpha(l,s)$ consisting of factors like in (950) such that

$$(952)\qquad \lambda(\overset{\circ}{l},Y,p) = \text{const}\ \alpha(l,s)\cdot\lambda(l,Y,s)\ .$$

Now let

$$(953)\qquad l_1 = \dots = l_w = m\ .$$

Then we set

$$(954)\qquad [m,w] = l = (m,\dots,m)\ ,$$

w times. Especially

$$(955)\qquad \overset{\circ}{l} = [1,n]\ .$$

From (953) we deduce

$$(956)\qquad n = mw\ .$$

THEOREM 86: It is

$$(957)\qquad \operatorname*{Res}_{u_1=m}\ \dots\ \operatorname*{Res}_{u_{w-1}=m}\ \zeta^*([m,w],Y,s) = (\text{Det } Y)^{-\frac{n^*}{2}}\ .$$

Especially

$$(958)\qquad \operatorname*{Res}_{u_1=1}\ \dots\ \operatorname*{Res}_{u_{n-1}=1}\ \zeta^*(\overset{\circ}{l},Y,s) = (\text{Det } Y)^{-\frac{n-1}{2}}\ .$$

PROOF: Apply theorem 84.

Set

(959) $$C(m,z) = \prod_{\iota=1}^{m} \{\overset{\circ}{\xi}(z + \tfrac{\iota}{2})(-z + \tfrac{\iota}{2})\} .$$

Then

(960) $$\Phi_o([m,w],s) = \Xi([m,w],s) = \prod_{1 \leq \mu < \nu \leq w} C(m, s_\nu - s_\mu) .$$

Let $F(z)$ be defined by (103), i. e.,

(961) $$F(z) = (1-z)\overset{\circ}{\xi}(z).$$

Then

(962) $$C(1,z) = F(z + \tfrac{1}{2}) .$$

Now let

(963) $$\overset{(m)}{s} = (s_1, s_1 + \tfrac{m}{2}, \ldots, s_1 + \tfrac{(w-1)m}{2}) .$$

Then

(964) $$\{1, \overset{(m)}{s}\} = s_1 + \frac{n^*}{4} .$$

Because of (959) the function $C(m,z)$ has exactly one zero of first order for $z = \frac{m}{2}$. Hence by theorem 86:

THEOREM 87: It is

(965) $$\lambda([m,w], Y, \overset{(m)}{s}) = \text{const}$$

with a constant not depending on Y and s_1.

THEOREM 88: The function $\lambda([m,w],Y,s)$ is invariant under all permutations of $s_1,\dots,s_w$.

PROOF: Apply (878), (936), (937), (939), (940) .

CHAPTER IV. SELBERG'S EISENSTEINSERIES

Selberg's Eisensteinseries are generalizations of Eisensteinseries for the elliptic modular group. They depend on several complex variables. Again we prove analytic continuation and functional equations. We proceed like in Diehl [11].

---

§ 14. SIEGEL'S UPPER HALF-PLANE

The real symplectic group $Sp(n, \mathbb{R})$ and Siegel's modular group $Sp(n, \mathbb{Z})$ operate on Siegel's upper half plane

$$\mathfrak{Z}(n) = \{Z = Z' = X + iY;\ Y > 0\}.$$

Some elementary results on Siegel's upper half-plane are collected.

The following results may be found for instance in Christian [7]. Siegel's upper half plane of degree n is given by

(966) $$\mathfrak{Z}(n) = \{Z = Z' = X + iY;\ Y \in \mathfrak{P}(n)\}\ .$$

It has

(967) $$d(n) = \frac{n(n+1)}{2}$$

complex dimensions. Form the $2n\times 2n$ matrix

(968) $$I = \begin{pmatrix} 0 & E \\ -E & 0 \end{pmatrix}$$

with $n\times n$ zero and identity matrices 0 and E. The real symplectic group $Sp(n, \mathbb{R})$ consists of all real $2n\times 2n$ matrices M with

(969) $$I[M] = I\ .$$

Set

(970) $$M = \begin{pmatrix} A & B \\ C & D \end{pmatrix}$$

with nxn matrices A, B, C, D. Then by

(971) $$M\langle Z\rangle = (AZ + B)(CZ + D)^{-1} = X_M + iY_M$$

the group $Sp(n, \mathbb{R})$ operates transitively on $\mathfrak{Z}(n)$. In (971) X_M and Y_M are the real and imaginary part of $M\langle Z\rangle$. We set

(972) $$M\{Z\} = CZ + D .$$

One has

(973) $$\operatorname{Det} M\{Z\} \neq 0 \qquad (M \in Sp(n, \mathbb{R})),$$

(974) $$Y_M = (M\{Z\})'^{-1} Y (M\{\overline{Z}\})^{-1} ,$$

hence

(975) $$\operatorname{Det} Y_M = (\operatorname{abs} M\{Z\})^{-2} \operatorname{Det} Y .$$

An easy computation shows

(976) $$(M_1M_2)\langle Z\rangle = M_1\langle M_2\langle Z\rangle\rangle \qquad (M_1, M_2 \in Sp(n, \mathbb{R})) ,$$

(977) $$(M_1M_2)\{Z\} = M_1\{M_2\langle Z\rangle\} M_2\{Z\} \quad (M_1, M_2 \in Sp(n, \mathbb{R})).$$

Furthermore one has $M\langle Z\rangle = Z$ for all $Z \in \mathfrak{Z}(n)$ if and only if $M = \pm E$.

The Jacobian of the substitution (971) is

(978) $$\mathfrak{J}_M(Z) = (\operatorname{Det} M\{Z\})^{-n-1} .$$

A volume element in $\mathfrak{Z}(n)$ is given by

(979) $$d\omega_Z = \frac{[dX][dY]}{(\operatorname{Det} Y)^{n+1}} ,$$

where $[dX]$, $[dY]$ denote the Euklidean volume elements in the X- and Y-spaces respectively. One has

(980) $$d\omega_{M\langle Z\rangle} = d\omega_Z \qquad (M \in Sp(n, \mathbb{R}))\ .$$

THEOREM 89: Let $\mathfrak{K} \subset \mathfrak{Z}(n)$ be a compact domain. Then there is a constant $c_{17} = c_{17}(n,\mathfrak{K}) \geq 1$ with

(981) $$Y_M \leq c_{17}\ Y_M^* \qquad (M \in Sp(n, \mathbb{R}))$$

for all $Z, Z^* \in \mathfrak{K}$.

PROOF: See Christian [6], page 271, Hilfssatz 15.

THEOREM 90: Let $\mathfrak{K} \subset \mathfrak{Z}(n)$ be a compact domain and c_{17} like in theorem 89. Then for

(982) $$Z, Z^* \in \mathfrak{K}\ ,$$

(983) $$M \in Sp(n, \mathbb{R})$$

the following inequalities hold

(984) $$c_{17}^{-1}(Y_M^*)_\nu \leq (Y_M)_\nu \leq c_{17}(Y_M^*)_\nu \quad (\nu = 1,\ldots,n)\ ,$$

(985) $$c_{17}^{-1}((Y_M^*)^{-1})_\nu \leq ((Y_M)^{-1})_\nu \leq c_{17}((Y_M^*)^{-1})_\nu \ (\nu=1,\ldots,n),$$

(986) $$c_{17}^{-\nu}\ \mathrm{Det}((Y_M^*)_\nu) \leq \mathrm{Det}((Y_M)_\nu) \leq c_{17}^{\nu}\ \mathrm{Det}((Y_M^*)_\nu)\ (\nu=1,\ldots,n),$$

(987) $$c_{17}^{-\nu}\mathrm{Det}(((Y_M^*)^{-1})_\nu) \leq \mathrm{Det}(((Y_M)^{-1})_\nu) \leq c_{17}^{\nu}\ \mathrm{Det}(((Y_M^*)^{-1})_\nu) \quad (\nu = 1,\ldots,n)\ .$$

PROOF: From (981) follows $Y_M^{-1} \leq c_{17} Y_M^{*-1}$. From this and (981) follow (984), (985) and from this (986), (987). Theorem 90 is proved.

DEFINITION 10: Let φ be a subgroup of $Sp(n, \mathbb{R})$ and

(988) $$1 \leq j \leq n \ .$$

Then let φ_j denote the subgroup of all $M \in \varphi$ of type

(989) $$M = \begin{pmatrix} A_j & 0 & B_j & * \\ * & P_j' & * & * \\ C_j & 0 & D_j & * \\ 0 & 0 & 0 & P_j^{-1} \end{pmatrix}$$

with a $j \times j$ matrix P_j and $(n-j)\times(n-j)$ matrices A_j, B_j, C_j, D_j. The group φ_j is called the "j-th cusp group" of φ.

DEFINITION 11: Siegel's modular group $\Gamma(n) = Sp(n, \mathbb{Z})$ consists of all integral matrices of $Sp(n, \mathbb{R})$. The Borelgroup

(990) $$\Phi(1) = \bigcap_{\nu=1}^{w} \Gamma_{k_\nu}(n)$$

consists of all matrices

(991) $$M = \begin{pmatrix} U' & SU^{-1} \\ 0 & U^{-1} \end{pmatrix}$$

with

(992) $$U \in \Delta(1)$$

and integral symmetric $S = S'$.

DEFINITION 12: Let $Y \in \mathcal{Y}(n)$ and set

(993) $$m(Y) = \min_{g \in \mathbb{Z}^n - \{0\}} Y[g] > 0 \ .$$

THEOREM 91: Let $\mathcal{K} \subset \mathcal{Z}(n)$ be a compact domain. There exists a constant $c_{18} = c_{18}(n, \mathcal{K}) \geq 1$ such that

(994) $\quad \mathrm{Det}(((Y_M)^{-1})_\nu) \geq c_{18}^{-1} \qquad (\nu = 1,\dots,n)$,

(995) $\quad \varkappa((Y_M)^{-1}) \geq c_{18}^{-1}$

for all $Z \in \tilde{\mathcal{K}}$ and $M \in \Gamma(n)$ holds.

PROOF: By enlarging $\tilde{\mathcal{K}}$ we may assume $iE \in \tilde{\mathcal{K}}$. Applying (985) with $Z^* = iE$ we see that it suffices to prove

(996) $\quad \mathrm{Det}(((E_M)^{-1})_\nu) \geq 1 \qquad (\nu = 1,\dots,n)$,

(997) $\quad \varkappa((E_M)^{-1}) \geq 1$.

But from (972), (974) we obtain

(998) $\quad (E_M)^{-1} = CC' + DD' > 0$.

This matrix is integral and hence (996), (997) follow. Theorem 91 is proved.

Set

(999) $\quad \check{W} = \begin{pmatrix} W(n) & 0 \\ 0 & W(n) \end{pmatrix} \in \Gamma(n)$.

Then the group

(1000) $\quad \check{\Phi}(\tilde{l}) = \check{W}\Phi(l)\check{W}$

consists of all matrices

(1001) $\quad M = \begin{pmatrix} U' & SU^{-1} \\ 0 & U^{-1} \end{pmatrix}$

with integral $S = S'$ and

(1002) $\quad U' \in \Delta(\tilde{l})$.

THEOREM 92: A fundamental domain $\mathcal{G}(n)$ of $\Gamma(n)$ in $\mathfrak{Z}(n)$ is given by

(1003) $\operatorname{abs} N\{Z\} \geq 1 \qquad (N \in \Gamma(n))$,

(1004) $Y \in \mathcal{M}(n)$,

(1005) $-\frac{1}{2} \leq x_{\iota\varkappa} \leq \frac{1}{2} \qquad (\iota,\varkappa = 1,\ldots,n)$

for

(1006) $X = (x_{\iota\varkappa})$.

PROOF: See Christian [7], Kapitel IV.3.

DEFINITION 13: Let $\mu > 0$. The "elementary set" $\mathcal{E}(n,\mu) \subset \mathfrak{Z}(n)$ is defined by

(1007) $Y \in \mathcal{Y}(n,\mu)$,

(1008) $y_1 > \mu^{-1}$,

(1009) $-\mu < x_{\iota\varkappa} < \mu \qquad (\iota,\varkappa = 1,\ldots,n)$.

THEOREM 93: There exists a constant $c_{19} = c_{19}(n) > 1$ such that

(1010) $\mathcal{G}(n) \subset \mathcal{E}(n,c_{19})$.

PROOF: See Christian [7], page 225, Satz 4.86.

THEOREM 94: There are only finite many $M \in \Gamma(n)$ such that

(1011) $M\langle\mathcal{E}(n,\mu)\rangle \cap \mathcal{E}(n,\mu) \neq \emptyset$.

PROOF: Christian [7], page 225, Satz 4.87.

THEOREM 95: Let $\mathcal{K} \subset \mathcal{Z}(n)$ be compact. Then there are only finite many $M \in \Gamma(n)$ such that

(1012) $M\langle \mathcal{K} \rangle \cap \mathcal{K} \neq \emptyset$

or

(1013) $M\langle \mathcal{K} \rangle \cap \mathcal{E}(n,\mu) \neq \emptyset$

or

(1014) $M\langle \mathcal{K} \rangle \cap \mathcal{G}(n) \neq \emptyset$.

PROOF: For sufficiently large μ^* we have $\mathcal{K}, \mathcal{G}(n), \mathcal{E}(n,\mu) \subset \mathcal{E}(n,\mu^*)$. Now theorem 95 follows from theorem 94.

§ 15. SELBERG'S EISENSTEINSERIES

Selberg's Eisensteinseries is defined and its convergence investigated.

DEFINITION 14: Let

(1015) $Z \in \mathcal{Z}(n)$,

(1016) $r \in \mathbb{N} \cup 0$

and form "Selberg's Eisensteinseries"

(1017) $$\mathcal{E}(l,r,Z,s) = \sum_{\Phi(l)\backslash\Gamma(n)} \operatorname{Det}(M\{\overline{Z}\})^{2r} f(l, Y_M, s+(r+\tfrac{n+1}{4})e(w)) .$$

DEFINITION 15: Let $\mathfrak{A} \subset \mathbb{R}^w$ be the domain

(1018) $$\sigma_1 > \frac{1+l_1}{4} ,$$

$$(1019)\qquad \sigma_{\nu+1} - \sigma_\nu > \frac{l_{\nu+1} + l_\nu}{4} \qquad (\nu = 1,\dots,w-1) .$$

Set

$$(1020)\qquad \mathfrak{b} = \mathfrak{a} + i\,\mathbb{R}^w .$$

THEOREM 96: The series $\mathcal{E}(l,r,Z,s)$ converges absolutely in $\mathfrak{b}$ and represents a holomorphic function there. If $\mathfrak{k} \subset \mathfrak{a}$, $\tilde{\mathfrak{k}} \subset \mathfrak{Z}(n)$ are compact subsets the series converges absolutely uniformly for

$$(1021)\qquad s \in \mathfrak{k} + i\,\mathbb{R}^w,\ Z \in \tilde{\mathfrak{k}} .$$

PROOF: From (975) we see

$$(1022)\qquad (\operatorname{abs} M\{\overline{Z}\})^{2r} = \frac{(\operatorname{Det} Y)^r}{(\operatorname{Det} Y_M)^r} .$$

Hence it suffices to prove that the series with positive terms

$$(1023)\qquad \mathcal{E}(l,0,Z,\sigma) = \sum_{\Phi(l)\backslash\Gamma(n)} f(l,Y_M,\sigma + \tfrac{n+1}{4}\,e(w))$$

converges uniformly for

$$(1024)\qquad \sigma \in \mathfrak{k}\ ,\ Z \in \tilde{\mathfrak{k}} .$$

Let $\mathcal{L} \subset \tilde{\mathfrak{k}}$ be a domain with

$$(1025)\qquad M\langle \mathcal{L}\rangle \cap \mathcal{L} = \emptyset \qquad (M \neq \pm E,\ M \in \Gamma(n)),$$

$$(1026)\qquad \int_{\mathcal{L}} d\omega_Z > 0 .$$

From (424), (433), (984), (1024), (1026) we deduce

$$(1027)\qquad f(l,Y_M,\sigma + \tfrac{n+1}{4}\,e(w)) \leq d_1 \int_{\mathcal{L}} f(\tilde{l},\tilde{Y}_M,\tilde{\sigma} - \tfrac{n+1}{4}\,e(w))\,d\omega_Z .$$

But

$$\check{\tilde{Y}}_M = (Y_{(\check{W}M)})^{-1} . \tag{1028}$$

In (1023) with M also $M\check{W}$ runs over $\Phi(1)\backslash\Gamma(n)$. Hence up to d_1 the sum (1023) is majorized by

$$H_1 = \sum_{\check{\Phi}(\tilde{1})\backslash\Gamma(n)} \int_{\mathcal{L}} f(\tilde{1},(Y_M)^{-1},\check{\sigma} - \frac{n+1}{4} e(w))d\omega_Z . \tag{1029}$$

Set

$$\mathfrak{h} = \bigcup_{M \in \check{\Phi}(\tilde{1})\backslash\Gamma(n)} M\langle \mathcal{L} \rangle . \tag{1030}$$

We may assume that $\mathfrak{h}$ lies in a fundamental domain of $\check{\Phi}(\tilde{1})$. Because of (1025) we have

$$H_1 = \int_{\mathfrak{h}} f(\tilde{1},Y^{-1},\check{\sigma} - \frac{n+1}{4} e(w))d\omega_Z . \tag{1031}$$

From (994) we deduce

$$\mathrm{Det}(Y^{-1})_\nu \geq c_{18}^{-1} \qquad (Z \in \mathfrak{h};\ \nu = 1,\ldots,n) \tag{1032}$$

Hence if H_1 converges in $\mathfrak{a}$ it converges uniformly for $\sigma \in \mathfrak{K}$. Therefore it is left to prove that H_1 converges for $\sigma \in \mathfrak{a}$.

From (355),(979) we deduce

$$d\omega_Z = (\mathrm{Det}\ Y)^{-\frac{n+1}{2}} [dX]dv_Y . \tag{1033}$$

Let $\mathfrak{f}$ be a fundamental domain of $\Delta(\tilde{1})$ in $\mathfrak{y}(n)$. Then a fundamental domain of $\check{\Phi}(\tilde{1})$ in $\mathfrak{z}(n)$ is given by

$$Y^{-1} \in \mathfrak{f} ;\ X = (x_{\iota\varkappa});\ -\frac{1}{2} \leq x_{\iota\varkappa} \leq \frac{1}{2} \quad (\iota,\varkappa=1,\ldots,n) . \tag{1034}$$

Hence by substituting $Y \to Y^{-1}$ we get

$$(1035)\qquad H_1 \leq \int\limits_{\substack{Y\in\mathcal{F}\\ \text{Det } Y_\nu \geq c_{18}^{-1}\ (\nu=1,\dots,n)}} f(\tilde{1},Y,\tilde{\sigma}+\tfrac{n+1}{4}\,e(w))\,dv_Y \quad .$$

Now we proceed like in the proof of theorem 61. We apply the generalized Jacobi trnasformation (379), (380), (381) with $\tilde{1}$ instead of 1. Instead of (718) we get

$$(1036)\qquad dv_Y = \prod_{\nu=1}^{w}\{(\text{Det } R_\nu)^{\frac{1}{2}(n-\tilde{k}_\nu-\tilde{k}_{\nu-1})}dv_{R_\nu}\}[dD] \; .$$

Furthermore

$$(1037)\qquad f(\tilde{1},Y,\tilde{\sigma}+\tfrac{n+1}{4}\,e(w)) = \prod_{\nu=1}^{w}(\text{Det } R_\nu)^{\tilde{\sigma}_\nu+\frac{1}{4}(\tilde{k}_{\nu-1}+\tilde{k}_\nu+1)} \; .$$

The fundamental domain $\mathcal{F}$ of $\Delta(\tilde{1})$ may be described by

$$(1038)\qquad R_\nu \in \mathcal{M}(\tilde{1}_\nu) \qquad\qquad (\nu = 1,\dots,w)$$

and the condition that all elements of D are between 0 and 1. Putting this into (1035) and immediately integrating over D gives

$$(1039)\quad H_1 \leq \int\limits_{\substack{R_\nu\in\mathcal{M}(\tilde{1}_\nu)\\ c_{18}^{-1}\leq\prod_{\mu=1}^{\nu}(\text{Det } R_\mu)}}^{(\nu=1,\dots,w)} \prod_{\nu=1}^{w}\{(\text{Det } R_\nu)^{\tilde{\sigma}_\nu+\frac{2n-\tilde{k}_\nu-\tilde{k}_{\nu-1}+1}{4}}\,dv_{R_\nu}\} \; .$$

Obviously

$$(1040)\qquad \tilde{\sigma}_\nu + \frac{2n-\tilde{k}_\nu-\tilde{k}_{\nu-1}+1}{4} = -\sigma_{w+1-\nu} + \frac{k_{w-\nu}+k_{w+1-\nu}+1}{4} < 0$$

because from (1018), (1019) we deduce

$$(1041)\qquad \sigma_\nu > \tfrac{1}{4}(k_\nu + k_{\nu-1} + 1) \; .$$

Hence we may apply (726) upon (1039) and we immediately see that up to a constant H_1 coincides with

$$(1042) \qquad H_2 = \int_{c_{18}^{-1} \leq \prod_{\mu=1}^{\nu} r_\mu} \prod_{\nu=1}^{w} \left(r_\nu^{\frac{k_{w-\nu}+k_{w+1-\nu}+1}{4} - \sigma_{w+1-\nu}} \frac{dr_\nu}{r_\nu}\right) .$$

Make the substitution (728), i. e.,

$$(1043) \qquad t_\nu = \prod_{\mu=1}^{\nu} r_\mu \qquad (\nu=1,\ldots,w) .$$

From (729) we deduce

$$(1044) \qquad \frac{dr_1}{r_1} \cdots \frac{dr_w}{r_w} = \frac{dt_1}{t_1} \cdots \frac{dt_w}{t_w} .$$

Furthermore

$$(1045) \qquad \prod_{\nu=1}^{w} r_\nu^{\frac{k_{w-\nu}+k_{w+1-\nu}+1}{4} - \sigma_{w+1-\nu}} = \prod_{\nu=1}^{w} t_\nu^{-g_\nu}$$

with

$$(1046) \qquad g_\nu = \sigma_{w+1-\nu} - \sigma_{w-\nu} - \frac{l_{w-\nu}+l_{w+1-\nu}}{4} \qquad (\nu = 1,\ldots,w-1) ,$$

$$(1047) \qquad g_w = \sigma_1 - \frac{l_1+1}{4} .$$

And from (1018), (1019) we get

$$(1048) \qquad g_\nu > 0 \qquad (\nu = 1,\ldots,w) .$$

Hence

$$(1049) \qquad H_2 = \prod_{\nu=1}^{w} \int_{c_{18}^{-1}}^{\infty} t_\nu^{-1-g_\nu} \, dt_\nu .$$

This converges. Theorem 96 is proved.

Let ψ be a row of principal characters. Then the series $E(1,r,Z,s)$ contains Selberg's zetafunction $\zeta^*(1,\psi,Y,s)$ as a partial sum. Hence $\zeta^*(1,\psi,Y,s)$ must converge in the domain (1018) (1019). But $\zeta^*(1,\psi,Y,s)$ does not change if one substitutes s by $s + ae(w)$ $(a \in \mathbb{C})$. Hence the condition (1018) can play no rôle in the convergence of $\zeta^*(1,\psi,Y,s)$. This leaves the conditions (1019) which are identical with (779). Hence we got a new proof for the convergence of Selberg's zetafunction in the domain (779).

An easy computatin shows

(1050) $\mathcal{E}(1,r,M\langle Z\rangle,s) = (\mathrm{Det}\ M\{\overline{Z}\})^{-2r}\,\mathcal{E}(1,r,Z,s) \quad (M \in \Gamma(n))$.

§ 16. REPRESENTATION WITH SIEGEL'S EISENSTEINSERIES

Like in Diehl [11] Selberg's Eisensteinseries is represented with the aid of Siegel's Eisensteinseries. This gives analytic continuation to a bigger domain and a certain functional equation.

From now on we assume

(1051) $l_1 = \dots = l_w = m$.

Then

(1052) $n = mw$.

We define Siegel's Eisensteinseries of degree m by

(1053) $$E(m,r,Z,u) = \sum_{M \in \Gamma_m(m)\backslash\Gamma(m)} (\mathrm{Det}\ M\{Z\})^{-2r}(\mathrm{Det}\ Y_M)^{u-r}$$

with a complex variable u.

THEOREM 97: For

$$\operatorname{Re} u > \frac{m+1}{2} \tag{1054}$$

the series (1053) converges absolutely and represents a holomorphic function in u.

PROOF: This follows easily from theorem 96 applied with $w = 1$ or from Braun [5].

From (990), (1051) we deduce

$$\Phi = \Phi([m,w]) = \bigcap_{\nu=1}^{w} \Gamma_{m_\nu}(n) . \tag{1055}$$

Set

$$\Phi^* = \Phi^*([m,w]) = \bigcap_{\nu=1}^{w-1} \Gamma_{m_\nu}(n) . \tag{1056}$$

Then

$$\breve{\Phi} = \breve{\Phi}([m,w]) = \breve{W}\, \Phi\, \breve{W} . \tag{1057}$$

Set

$$\breve{\Phi}^* = \breve{W}\, \Phi^*\, \breve{W} . \tag{1058}$$

Obviously

$$\Phi \subset \Phi^*; \ \breve{\Phi} \subset \breve{\Phi}^* . \tag{1059}$$

THEOREM 98: Let

$$\hat{J} = \begin{pmatrix} A & B \\ C & D \end{pmatrix} \in \Gamma(m) \tag{1060}$$

run over $\Gamma_m(m)\backslash\Gamma(m)$. Then

$$(1061) \qquad \mathring{J} = \begin{pmatrix} A & 0 & B & 0 \\ 0 & E^{(n^*)} & 0 & 0 \\ C & 0 & D & 0 \\ 0 & 0 & 0 & E^{(n^*)} \end{pmatrix}$$

runs over a complete set of representatives of $\Phi\backslash\Phi^*$ and

$$(1062) \qquad J = \begin{pmatrix} E^{(n^*)} & 0 & 0 & 0 \\ 0 & A & 0 & B \\ 0 & 0 & E^{(n^*)} & 0 \\ 0 & C & 0 & D \end{pmatrix}$$

runs over a complete set of representatives of $\check{\Phi}\backslash\check{\Phi}^*$.

PROOF: It suffices to prove the first assertion. The matrices $N \in \Phi^*$ are of type

$$(1063) \qquad N = \begin{pmatrix} A & 0 & B & * \\ * & P' & * & * \\ C & 0 & D & * \\ 0 & 0 & 0 & P^{-1} \end{pmatrix}$$

with (1060) and

$$(1064) \qquad P \in \Delta([m,w-1]) \ .$$

Hence

$$(1065) \qquad N = K\,\mathring{J}$$

with

$$(1066) \qquad K \in \Phi \ .$$

Therefore each residue class of $\Phi\backslash\Phi^*$ contains a representative of type (1061).

Now let $\mathring{J}_1$, $\mathring{J}_2$ be two matrices of type (1061) and $\mathring{J}_2 = K\mathring{J}_1$ with (1066). Then one easily computes

$$(1067) \qquad K = \begin{pmatrix} K_1' & 0 & K_2K_1^{-1} & 0 \\ 0 & E^{(n^*)} & 0 & 0 \\ 0 & 0 & K_1^{-1} & 0 \\ 0 & 0 & 0 & E^{(n^*)} \end{pmatrix}$$

with

$$(1068) \qquad \begin{pmatrix} K_1' & K_2K_1^{-1} \\ 0 & K_1^{-1} \end{pmatrix} \in \Gamma_m(m) .$$

This proves the theorem

THEOREM 99: Let

$$(1069) \qquad Z = \begin{pmatrix} Z_1^* & Z_2^* \\ Z_2^{*\prime} & \hat{Z} \end{pmatrix} \in \mathfrak{Z}(n)$$

with

$$(1070) \qquad \hat{Z} \in \mathfrak{Z}(m) .$$

Hence $\hat{Z}$ is the right lower $m\times m$ submatrix of Z. Set

$$(1071) \qquad \hat{Z} = \hat{X} + i\hat{Y} .$$

Then

$$(1072) \qquad \mathcal{E}([m,w],r,Z,s) =$$

$$\sum_{M \in \check{\Phi}^*\backslash\Gamma(n)} (\operatorname{Det} \hat{Y}_M)^{2r}(\operatorname{Det} M\{\overline{Z}\})^{2r} f([m,w-1],(Y_M^{-1})_{n^*},(\tilde{s})^*-(r + \tfrac{n+m+1}{4})e(w^*))$$

$$\times \qquad E(m,r,\widehat{M\langle Z\rangle},s_1 + \tfrac{m+1}{4}) .$$

PROOF: From (433), (1017) we deduce

$$(1073) \quad \mathcal{E}([m,w],r,Z,s) = \sum_{M \in \Phi\backslash\Gamma(n)} (\operatorname{Det} M\{\overline{Z}\})^{2r} f([m,w],\tilde{Y}_M,\tilde{s}-(r+ \tfrac{n+1}{4})e(w)) .$$

From (977), (999) we get

$$(1074) \qquad \mathrm{Det}((\breve{W}M)\{Z\}) = \mathrm{Det}(M\{Z\}) \ .$$

Furthermore we have (1028). With M also $M\breve{W}$ runs over $\Phi\backslash\Gamma(n)$. Writing M instead of $\breve{W}M\breve{W}$ we get

$$(1075) \qquad \mathcal{E}([m,w],r,Z,s) =$$

$$\sum_{M \in \breve{\Phi}\backslash\Gamma(n)} (\mathrm{Det}\ M\{\overline{Z}\})^{2r} f([m,w],Y_M^{-1},\tilde{s} - (r + \tfrac{n+1}{4})e(w)) \ .$$

Hence

$$(1076) \qquad \mathcal{E}([m,w],r,Z,s) =$$

$$\sum_{M \in \breve{\Phi}^*\backslash\Gamma(n)} \ \sum_{J \in \breve{\Phi}\backslash\breve{\Phi}^*} (\mathrm{Det}(JM)\{\overline{Z}\})^{2r} f([m,w],Y_{JM}^{-1},\tilde{s} - (r + \tfrac{n+1}{4})e(w)) \ .$$

Here we assume that J is of type (1063).

Now

$$(1077) \qquad \mathrm{Det}((JM)\{\overline{Z}\}) = (\mathrm{Det}\ M\{\overline{Z}\})(\mathrm{Det}\ \hat{J}\{\widehat{M\langle\overline{Z}\rangle}\}) \ ,$$

hence

$$(1078) \qquad \mathrm{Det}((JM)\{\overline{Z}\}) = (\mathrm{Det}\ M\{\overline{Z}\})(\mathrm{Det}\ \hat{Y}_M)(\mathrm{Det}(\hat{Y}_M)_{\hat{J}})^{-1}(\mathrm{Det}\ \hat{J}\{\widehat{M\langle Z\rangle}\})^{-1} .$$

Furthermore

$$(1079) \qquad (Y_{JM})^{-1} = ((Y_M)_J)^{-1} \ ,$$

$$(1080) \qquad ((Y_{JM})^{-1})_{n*} = (Y_M^{-1})_{n*} \ ,$$

$$(1081) \qquad \mathrm{Det}(Y_{JM}) = (\mathrm{Det}\ Y_M)\mathrm{abs}(\hat{J}\{\widehat{M\langle Z\rangle}\})^{-2} \ .$$

From (434) we deduce

(1082) $$\operatorname{Det} Y_M = (\operatorname{Det} \widehat{Y_M})(\operatorname{Det}((Y_M^{-1})_{n*}))^{-1} .$$

From (1081), (1082) we get

(1083) $$\operatorname{Det}(Y_{JM}) = (\operatorname{Det}((Y_M^{-1})_{n*}))^{-1}(\operatorname{Det}(\widehat{Y_M})_{\hat{J}}) .$$

From (427), (1080) we deduce

(1084) $$f([m,w],(Y_{JM})^{-1},\tilde{s} - (r + \tfrac{n+1}{4})e(w)) =$$

$$(\operatorname{Det} Y_{JM})^{s_1+(r + \frac{m+1}{4})} f([m,w-1],(Y_M^{-1})_{n*},(\tilde{s})^* + (s_1 - \tfrac{n}{4})e(w*)) .$$

The formulas (426), (1083), (1084) give us

(1085) $$f([m,w],(Y_{JM})^{-1},\tilde{s} - (r + \tfrac{n+1}{4})e(w)) =$$

$$f([m,w-1](Y_M^{-1})_{n*},(\tilde{s})^*-(r+ \tfrac{n+m+1}{4})e(w*))\operatorname{Det}((\widehat{Y_M})_{\hat{J}})^{s_1+r+ \frac{m+1}{4}} .$$

From (1076), (1078), (1085) we get

(1086) $$\mathcal{E}([m,w],r,Z,s) =$$

$$\sum_{M \in \check{\Phi}^*\backslash\Gamma(n)}(\operatorname{Det} \widehat{Y_M})^{2r}(\operatorname{Det} M\{\overline{Z}\})^{2r}f([m,w-1],(Y_M^{-1})_{n*},(\tilde{s})^*-(r+ \tfrac{n+m+1}{4})e(w*)) \times$$

$$\sum_{\hat{J} \in \Gamma_m(m)\backslash\Gamma(m)}(\operatorname{Det} \hat{J}\{\widehat{M\langle Z\rangle}\})^{-2r}(\operatorname{Det}(\widehat{Y_M})_{\hat{J}})^{s_1-r+ \frac{m+1}{4}} .$$

From (1053), (1086) we get (1072). Theorem 99 is proved.

Axiom A: There exists a function $K(m,r,u)$ with the following properties:

a) $K(m,r,u)$ is meromorphic for $u \in \mathbb{C}$ and possesses only finitely many poles.

b) The function

$$T(m,r,Z,u) = K(m,r,u)E(m,r,Z,u) \tag{1087}$$

is holomorphic for $u \in \mathbb{C}$ and satisfies the functional equation

$$T(m,r,Z,u) = T(m,r,Z,\frac{m+1}{2} - u) \ . \tag{1088}$$

c) There exist a constant $c_{19} \geq 1$, a certain natural number h and linear functions

$$\mathcal{L}(\iota,\mathrm{Re}\ u) = j^*(\iota)\mathrm{Re}\ u + j(\iota) \qquad (\iota = 1,\ldots,h) \tag{1089}$$

with

$$j^*(\iota) = -1,0,1 \tag{1090}$$

such that

$$\mathrm{abs}\ T(m,r,Z,u) \leq c_{19} \sum_{\iota=1}^{h} (\mathrm{Det}\ Y)^{\mathcal{L}(\iota,\mathrm{Re}\ u)} \quad (Z \in \mathcal{G}(m)) \ . \tag{1091}$$

THEOREM 100: For $m = 1$ Axiom A is true with

$$K(1,r,u) = \begin{Bmatrix} F(s) & (r = 0) \\ \xi(r,s) & (r > 0) \end{Bmatrix} . \tag{1092}$$

PROOF: See § 3.

DEFINITION 16: A domain $\mathcal{T} \subset \mathbb{R}^W$ is called a "$\mathcal{T}$-domain" if there exists a constant d with

(1093) $$\sigma_{\nu+1} - \sigma_\nu > d \qquad (\nu = 1,\dots,w-1),$$

(1094) $$\sigma_2 + \sigma_1 > d \quad ,$$

A domain $\vartheta \in \mathbb{C}^w$ is called a "ϑ-domain" if

(1095) $$\vartheta = \mathcal{t} + i\,\mathbb{R}^w$$

with a $\mathcal{t}$-domain $\mathcal{t}$.

THEOREM 101: A $\mathcal{t}$-domain is invariant under the substitution

(1096) $$\sigma_1 \to -\sigma_1,\ \sigma_\nu \to \sigma_\nu \qquad (\nu = 2,\dots,w) .$$

A ϑ-domain is invariant under the substitution

(1097) $$s_1 \to -s_1;\ s_\nu \to s_\nu \qquad (\nu = 2,\dots,w) .$$

PROOF: Clear

THEOREM 102: Let axiom A be true. The series

(1098) $$S(m,w,r,Z,s) = K(m,r,s_1 + \tfrac{m+1}{4})\,\mathcal{E}([m,w],r,Z,s) =$$

$$\sum_{M \in \check{\Phi}^* \backslash \Gamma(n)} (\mathrm{Det}\ \widehat{Y_M})^{2r}(\mathrm{Det}\ M\{\overline{Z}\})^{2r} f([m,w-1],(Y_M^{-1})_{n*},(\breve{s})^* - (r+ \tfrac{n+m+1}{4})e(w^*)) \times$$

$$T(m,r,\widehat{M\langle Z\rangle},s_1 + \tfrac{m+1}{4})$$

converges absolutely in a ϑ-domain ϑ. In ϑ it is holomorphic and invariant under the substitution (1097).

PROOF: The invariance under (1097) follows from (1088). Like in theorem 96 it suffices to prove the absolute convergence in a ϑ-domain. By multiplying M with a left-hand-factor $\in \check{\Phi}^*$ we may obtain

(1099) $$\widehat{M\langle Z\rangle} \in \mathcal{G}(m) .$$

Applying formula (1022) we see that it suffices to show that h sums of type

$$(1100) \quad \sum_{M \in \check{\Phi}^*\backslash\Gamma(n)} (\mathrm{Det}\ Y_M^{-1})^r f[m,w-1],(Y_M^{-1})_{n*},(\tilde{\sigma})^* - (r+ \frac{n+m+1}{4})\ e(w^*)) \quad \times$$

$$(\mathrm{Det}\ \hat{Y}_M)^{\mathcal{L}_1(\iota,\sigma_1)} \quad (\iota=1,\ldots,h)$$

converge with $\mathcal{L}_1(\iota,\sigma_1) = \mathcal{L}(\iota,\sigma_1) + 2r$.

Like in the proof of theorem 96 the sum converges if the integral

(1101) $\quad H_1 =$

$$\int_{\mathfrak{h}} (\mathrm{Det}\ Y^{-1})^r f([m,w-1],(Y^{-1})_{n*},(\tilde{\sigma})^* - (r+ \frac{n+m+1}{4}) e(w^*))(\mathrm{Det}\ \hat{Y})^{\mathcal{L}_1(\iota,\sigma_1)} d\omega_Z$$

converges. But now we may assume that $\mathfrak{h}$ lies in a fundamental domain of $\check{\Phi}^*$. Hence we get

$$(1102) \qquad \mathrm{Det}\ \hat{Y} \geq d_1^{-1}$$

with some constant $d_1 \geq 1$. Then instead of (1035) we obtain

(1103) $\quad H_1 \leq$

$$\int_{\substack{Y \in \mathfrak{f} \\ \mathrm{Det}\ Y_{k_\nu} \geq c_{18}^{-1}\ (\nu=1,\ldots,w) \\ \mathrm{Det}\ \widehat{Y^{-1}} \geq d_1^{-1}}} (\mathrm{Det}\ Y)^r f([m,w-1],Y_{n*},(\tilde{\sigma})^* - (r+ \frac{n+m+1}{4}) e(w^*))(\mathrm{Det}\ \widehat{Y^{-1}})^{\mathcal{L}_1(\iota,\sigma_1)} d\omega_Z .$$

Now apply again the generalized Jacobi transformation. Then one gets

$$(1104) \qquad \widehat{Y^{-1}} = R_w^{-1} .$$

Hence instead of (1042) we get

(1105) $$H_2 = \int\limits_{c_{18}^{-1} \leq \prod_{\mu=1}^{\nu} r_\mu \ (\nu=1,\ldots,w-1)} \prod_{\mu=1}^{w-1} r_\mu^{\frac{(w-\mu)m}{2} + \frac{m-1}{4} - \sigma_{w+1-\mu}} \times$$

$$\left(\int\limits_{c_{18}^{-1} r_1^{-1} \ldots r_{w-1}^{-1}}^{d_1} r_w^{\mathcal{L}_2(\iota,\sigma_1)} dr_w \right) \frac{dr_1}{r_1} \ldots \frac{dr_{w-1}}{r_{w-1}} \qquad (\iota = 1,\ldots,h)$$

with linear functions $\mathcal{L}_2(\iota,\sigma_1)$ of type (1089), (1090). The inner integral may be estimated by a sum of 3 terms of type

$$(\mathrm{const})^{\mathcal{L}_3(\varkappa,\sigma_1)} (r_1 \ldots r_{w-1})^{\mathcal{L}_3(\varkappa,\sigma_1) - \frac{m-1}{4}}$$

with linear functions $\mathcal{L}_3(\varkappa,\sigma_1)$ of type (1089), (1090). Hence it suffices to compute the integrals

(1106) $$\int\limits_{c_{18}^{-1} \leq \prod_{\mu=1}^{\nu} r_\mu \ (\nu=1,\ldots,w-1)} r_\nu^{\frac{(w-\nu)m}{2} + \mathcal{L}_3(\varkappa,\sigma_1) - \sigma_{w+1-\nu}} \frac{dr_1}{r_1} \ldots \frac{dr_{w-1}}{r_{w-1}} \quad (\varkappa=1,\ldots,3h).$$

Make again the substitution

(1107) $$t_\nu = \prod_{\mu=1}^{\nu} r_\mu \qquad (\nu = 1,\ldots,w-1) .$$

Then

(1108) $$\frac{dr_1}{r_1} \ldots \frac{dr_{w-1}}{r_{w-1}} = \frac{dt_1}{t_1} \ldots \frac{dt_{w-1}}{t_{w-1}} ,$$

(1109) $$\prod_{\nu=1}^{w-1} r_\nu^{\frac{(w-\nu)m}{2} + \mathcal{L}_3(\varkappa,\sigma_1) - \sigma_{w+1-\nu}} = \prod_{\nu=1}^{w-1} t_\nu^{-g_\nu}$$

with

(1110) $$g_\nu = \sigma_{w+1-\nu} - \sigma_{w-\nu} - \frac{m}{2} \qquad (\nu=1,\dots,w-2) \ ,$$

(1111) $$g_{w-1} = \sigma_2 - \mathcal{L}_3(\varkappa,\sigma_1) \qquad (\varkappa=1,\dots,3h) \ .$$

Then (1107) becomes

(1112) $$\prod_{\nu=1}^{w-1} \int_{c_{18}^{-1}}^{\infty} t_\nu^{-1-g_\nu} \, dt_\nu \ .$$

This converges for $g_1,\dots,g_{w-1} > 0$ which is identical with (1093), (1094) and

(1113) $$\sigma_2 > d$$

with a suitable constant d. But because of

(1114) $$\sigma_2 = \frac{\sigma_2+\sigma_1}{2} + \frac{\sigma_2-\sigma_1}{2}$$

the condition (1113) is a consequence of (1093), (1094) with another d. Hence we may confine to (1093), (1094). Theorem 102 is proved.

THEOREM 103: The poles of the function $C(m,u)$ are of first order. They lie in the points

(1115) $$z = -\frac{\iota}{2} \qquad (\iota = 0,1,\dots,m-1) \ .$$

PROOF: Apply (959) .

THEOREM 104: Let axiom A be true. The function

(1116) $$P(m,w,r,Z,s) = \Big(\prod_{\nu=1}^{w} K(m,r,s_\nu + \tfrac{m+1}{4})\Big) \times$$

$$\Big(\prod_{1 \le \mu < \nu \le w} (C(m,s_\nu - s_\mu)C(m,s_\nu + s_\mu))\Big) \mathcal{E}([m,w],r,Z,s)$$

is holomorphic in a ϑ-domain ϑ and invariant under (1098).

PROOF: For

$$(1117) \qquad K(m,r,s_1+\frac{m+1}{4})\mathcal{E}([m,w],r,Z,s)$$

this was proved in theorem 102. For $2 \leq \nu \leq w$ the function

$$(1118) \qquad K(m,r,s_\nu+\frac{m+1}{4})$$

is independent of s_1. From (1114) and (1094) we deduce

$$(1119) \qquad \sigma_\nu > d \qquad (\nu = 2,\ldots,w).$$

Since K has only finitely many poles we may choose d so large that (1118) is holomorphic in ϑ.

It is easily seen that the product

$$(1120) \qquad \prod_{1 \leq \mu < \nu \leq w} (C(m,s_\nu-s_\mu)C(m,s_\nu+s_\mu))$$

is invariant under (1097). From theorem 103 and formulas (1093), (1119) we see that it is holomorphic in ϑ. Theorem 104 is proved.

§ 17. REPRESENTATION WITH SELBERG'S ZETAFUNCTION

Like in Diehl [11] Selberg's Eisensteinseries is represented with the aid of Selberg's zetafunction. This gives analytic continuation to a bigger domain and the invariance under the permutations of all variables.

THEOREM 105: It is

$$(1121) \quad \mathcal{E}([m,w],r,Z,s)=\sum_{M\in\Gamma_n(n)\backslash\Gamma(n)}(\mathrm{Det}\ M\{\overline{Z}\})^{2r}(\mathrm{Det}\ Y_M)^{r+\frac{n+1}{4}}\Lambda([m,w],Y_M,s)\ .$$

PROOF: Apply (918), (1017).

We set

$$(1122) \qquad \{s\} = \{[m,w],s\} = \frac{1}{w}\sum_{\nu=1}^{w} s_\nu \quad .$$

Then from (933) we get

$$(1123) \qquad \Phi_0([m,w],s)\Lambda([m,w],Y,s) = (\operatorname{Det} Y)^{\{s\}}\lambda([m,w],Y,s) \quad .$$

From (1121), (1123) we deduce

$$(1124) \qquad \Phi_0([m,w],s)\,\mathcal{E}([m,w],r,Z,s) =$$

$$\sum_{M \in \Gamma_n(n)\backslash\Gamma(n)} (\operatorname{Det} M\{\overline{Z}\})^{2r}(\operatorname{Det} Y_M)^{\{s\}+r+\frac{n+1}{4}}\lambda([m,w],Y_M,s) \quad .$$

DEFINITION 17: A domain $\mathcal{R} \subset \mathbb{R}^w$ is called an "$\mathcal{R}$-domain", if the following holds: There are finitely many (say a) linear functions $\mathcal{L}(\varkappa,\dots)$ ($\varkappa = 1,\dots,a$) such that

$$(1125) \quad \sigma_\nu \geq \mathcal{L}(\varkappa;\sigma_\nu-\sigma_{\varphi_\nu(1)},\dots,\sigma_\nu-\sigma_{\varphi_\nu(\nu-1)},\sigma_\nu-\sigma_{\varphi_\nu(\nu+1)},\dots,\sigma_\nu-\sigma_{\varphi_\nu(w)}) \quad .$$

Here φ_ν runs over all permutations of $1,\dots,\nu-1,\nu+1,\dots,w$; $\nu = 1,\dots,w$; $\varkappa = 1,\dots,a$.

A domain $\mathcal{G} \subset \mathbb{C}^w$ is called a "$\mathcal{G}$-domain" if

$$(1126) \qquad \mathcal{G} = \mathcal{R} + i\,\mathbb{R}^w$$

with an $\mathcal{R}$-domain $\mathcal{R}$.

THEOREM 106: Each $\mathcal{R}$-domain is invariant under all permutaions of $\sigma_1,\dots,\sigma_w$. Each $\mathcal{G}$-domain is invariant under all permutations of $s_1,\dots,s_w$.

PROOF: Clear.

THEOREM 107: The function $\lambda(1,Y,s)$ depends only on the differences $s_\nu - s_\mu$ $(1 \leq \mu < \nu \leq w)$.

PROOF: Apply formula (947), i. e.,

$$\lambda(1,Y,s) = \Phi_o(1,s)(\operatorname{Det} Y)^{s_w - \{1,s\} + \frac{n^*}{4}} \zeta^*(1,Y,s) . \tag{1127}$$

Because of (927), (928), (930) we see that $\Phi_o(1,s)$ depends only on the differences $s_\nu - s_\mu$ $(1 \leq \mu < \nu \leq w)$. The same is true for $s_w - \{1,s\}$. And it is true for $\zeta^*(1,Y,s) = \zeta^*(1,Y,u)$. Theorem 107 is proved.

THEOREM 108: The series

$$Q(m,w,r,Z,s) = \Phi_o([m,w],s)\mathcal{E}([m,w],r,Z,s) = \tag{1128}$$

$$\sum_{M \in \Gamma_n(n)\backslash\Gamma(n)} (\operatorname{Det} M\{\overline{Z}\})^{2r}(\operatorname{Det} Y_M)^{\{s\}+r+\frac{n+1}{4}} \lambda([m,w],Y_M,s)$$

converges absolutely in a $\mathcal{Y}$-domain $\mathcal{Y}$. In $\mathcal{Y}$ it is holomorphic and invariant under all permutations of $s_1,\dots,s_w$.

PROOF: The invariance under all permutations of $s_1,\dots,s_w$ follows from theorem 88. Like in theorem 96 it suffices to prove the absolute convergence in a $\mathcal{Y}$-domain. Applying formula (1022) we see that it suffices to consider the case $r = 0$. Furthermore we may assume in (1128) that

$$Y_M \in \mathfrak{M}(n) \subset \mathcal{Y}(n,c_{11}) . \tag{1129}$$

Now we apply the estimate (844), i. e.,

$$\operatorname{abs} \lambda([m,w],Y,s) \leq c_{15}(\operatorname{Det} Y)^{\sigma_w - \{\sigma\} + \frac{n^*}{4}} \sum_{\iota=1}^{g(1)} y_1^{\mathcal{L}(1,\iota,\sigma)} + \tag{1130}$$

$$c_{16}(\operatorname{Det} Y)^{\sigma_w - \{\sigma\} - \frac{n^*}{4}} \sum_{\iota=1}^{g(1)} y_n^{-\mathcal{L}(\check{1},\iota,\check{\sigma})}.$$

From theorem 107 we deduce, that the linear functions $\mathcal{L}$ may be chosen in such a way that they depend only on the differences $\sigma_\nu - \sigma_\mu$ $(1 \leq \mu < \nu \leq w)$. Furthermore

$$(\mathrm{Det}\ Y)^{\pm \frac{n^*}{4} - \frac{n+1}{2}}$$

and $y_1^\gamma\, y_n^\delta$ may be both estimated by a finite sum of suitable powers of y_1 and y_n. Hence it suffices to prove the convergence of finitely many sums of type

$$\sum_{M \in \Gamma_n(n)\backslash\Gamma(n)} (\mathrm{Det}\ Y_M)^{\sigma_w + \frac{n+1}{2}}\, (y_M)_\beta^{\mathcal{B}(\varkappa)} \qquad (\beta=1,n;\ \varkappa=1,\dots,h)\ , \tag{1131}$$

where $\mathcal{B}(\varkappa)$ are linear functions in $\sigma_\nu - \sigma_\mu$ $(1 \leq \mu < \nu \leq w)$, provided that σ lies in a suitable $\mathfrak{n}$-domain $\mathfrak{n}$.

Like in the proof of theorem 96 the sum (1131) converges when the integral

$$H_1 = \int_{\mathfrak{z}} (\mathrm{Det}\ Y)^{\sigma_w + \frac{n+1}{2}}\, y_\beta^{\mathcal{B}(\varkappa)}\, d\omega_Z \qquad (\beta=1,n;\ \varkappa=1,\dots,h) \tag{1132}$$

converges. But now we may assume that $\mathfrak{z}$ is in a fundamental domain of $\Gamma_n(n)$. By theorem 91 we have

$$\mathfrak{m}(Y^{-1}) \geq c_{18}^{-1}\ . \tag{1133}$$

But $Y \in \mathfrak{M}(n)$ and hence (1133) is equivalent with

$$y_n \leq d_1 \tag{1134}$$

with some positive constant d_1. Apply (1034) and integrate over X. Then we get

$$H_1 \leq \int_{\substack{Y \in \mathfrak{M}(n) \\ y_n \leq d_1}} (\mathrm{Det}\ Y)^{\sigma_w}\, y_\beta^{\mathcal{B}(\varkappa)}\, dv_Y \qquad (\beta = 1,n;\ \varkappa = 1,\dots,h)\ . \tag{1135}$$

Now apply the Jacobi transformation (357), (358), (359). Then from (385) we get

$$(1136) \qquad dv_Y = \prod_{\nu=1}^{n} \left(r_\nu^{\frac{1}{2}(n-2\nu+1)} \frac{dr_\nu}{r_\nu} \right) [dD] \, .$$

Applying this to (1134) and immediately integrating over D we get up to a constant factor, the following two types of integrals:

$$(1137) \qquad \int\limits_{\substack{\frac{r_\nu}{r_{\nu+1}} \leq d_2 \ (\nu=1,\dots,n-1) \\ r_n \leq d_2}} r_n^{\mathcal{L}(\varkappa)} \left(\prod_{\nu=1}^{n} r_\nu^{\sigma_w + \frac{1}{2}(n-2\nu+1)} \frac{dr_\nu}{r_\nu} \right) \qquad (\varkappa = 1,\dots,h) \, ,$$

$$(1138) \qquad \int\limits_{\substack{\frac{r_\nu}{r_{\nu+1}} \leq d_2 \ (\nu=1,\dots,n-1) \\ r_n \leq d_2}} r_1^{\mathcal{L}(\varkappa)} \left(\prod_{\nu=1}^{n} r_\nu^{\sigma_w + \frac{1}{2}(n-2\nu+1)} \frac{dr_\nu}{r_\nu} \right) \qquad (\varkappa = 1,\dots,h) \, .$$

From

$$(1139) \qquad \frac{r_\nu}{r_{\nu+1}} \leq d_2 \qquad (\nu = 1,\dots,n-1) \, ,$$

$$(1140) \qquad r_n \leq d_2$$

we get

$$(1141) \qquad d_2^{1-n} r_1 \leq r_n \leq d_2 = d_2 \, r_1^o \, .$$

Hence the integral (1137) may be majorized by integrals of type (1138).

The integral (1138) has the upper estimate

$$(1142)\quad \Big(\int_0^{d_3} r_1^{\sigma_w + \frac{1}{2}(n-1) + \mathscr{B}(\varkappa)} \frac{dr_1}{r_1}\Big) \prod_{\nu=2}^{n} \int_0^{d_3} r_\nu^{\sigma_w + \frac{1}{2}(n-2\nu+1)} \frac{dr_\nu}{r_\nu}$$

$(\varkappa=1,\ldots,h)$

with a certain constant $d_3 > 0$. These integrals converge for

$$(1143)\qquad \sigma_w \geq -1 - \mathscr{B}(\varkappa) \qquad\qquad (\varkappa = 1,\ldots,h)\ ,$$

$$(1144)\qquad \sigma_w \geq \frac{n}{2} \quad .$$

On the right-hand-side of (1143), (1144) stand linear functions in $\sigma_\nu - \sigma_\mu$ $(1 \leq \mu < \nu \leq w)$. Ajoining some more inequalities which may make the domain smaller we see that we have convergence in a $\mathscr{Y}$-domain. Theorem 108 is proved.

THEOREM 109: Let axiom A be true. The function (1117) is holomorphic in a $\mathscr{Y}$-domain and invariant under all permutations of $s_1,\ldots,s_w$.

PROOF: Because of (960) and theorem 108 it suffices to show that the functions

$$(1145)\qquad \prod_{\nu=1}^{w} K(m,r,s_\nu + \frac{m+1}{4})\ ,$$

$$(1146)\qquad \prod_{1 \leq \mu < \nu \leq w} C(m,s_\nu + s_\mu)$$

are holomorphic in a $\mathscr{Y}$-domain. The invariance of (1145), (1146) under all permutations of $s_1,\ldots,s_w$ is immediately seen. Now assume

$$(1147)\qquad \sigma_\nu \geq d \qquad\qquad (\nu = 1,\ldots,w)$$

with a proper constant d. Then the holomorphy of (1146) follows from theorem 103 and the holomorphy of (1145) follows because K has only finite many poles. Ajoining (1147) to the conditions (1125) the assertion follows.

§ 18. ANALYTIC CONTINUATION

We multiply Selberg's Eisensteinseries with certain factors. The product is analytically continued to the whole complex space. It is invariant under a certain finite group.

THEOREM 110: Let $w \geq 2$ and let $\mathcal{R} \subset \mathbb{R}^w$ be a connected domain. If a function is holomorphic in $\mathcal{R} + i\,\mathbb{R}^w$ then it may be holomorphically extended to $\mathcal{f}(\mathcal{R}) + i\,\mathbb{R}^w$, where $\mathcal{f}(\mathcal{R})$ is the convex hull of $\mathcal{R}$.

PROOF: See Hörmander [18], page 41, theorem 2.5.10.

DEFINITION 18: Let $\Delta(w)$ denote the group of all substitutions

(1148) $$s_\nu \to \varepsilon_\nu\, s_{\varphi(\nu)} \qquad (\nu = 1,\ldots,w)\ .$$

Here φ runs over the permutations of $s_1,\ldots,s_w$. Furthermore $\varepsilon_1,\ldots,\varepsilon_w$ run independently over ± 1.

$\Delta(n)$ is exactly the group of all integral orthogonal substitutions. Its order is

(1149) $$2^n n!\ .$$

THEOREM 111: Let Axiom A be true. Then the function $P(m,w,r,Z,s)$ may be holomorphically continued to $\mathbb{C}^w$. It is invariant under $\Delta(w)$.

PROOF: Let $\mathcal{t}$ be a $\mathcal{t}$-domain and $\mathcal{w}$ an $\mathcal{w}$-domain. We show that

(1150) $$\mathcal{t} \cap \mathcal{w} \neq \emptyset\ .$$

The domain $\mathcal{t}$ is given by (1093), (1094), i. e.,

(1151) $$\sigma_{\nu+1} - \sigma_\nu > d \qquad (\nu = 1,\ldots,w-1),$$

(1152) $$\sigma_2 + \sigma_1 > d .$$

The domain $\mathcal{W}$ is given by (1125), i. e.,

(1153) $$\sigma_\nu > \mathcal{L}(\varkappa;\sigma_\nu-\sigma_{\varphi(1)},\dots,\sigma_\nu-\sigma_{\varphi(\nu-1)},\sigma_\nu-\sigma_{\varphi(\nu+1)},\dots,\sigma_\nu-\sigma_{\varphi(w)})$$

with linear functions $\mathcal{L}$. φ runs over $\overset{(\nu)}{\mathcal{Y}}(w^*)$ where $\mathcal{Y}(w^*)$ denotes the symmetric group of degree w^* and (ν) means that it operates on $1,\dots,\nu-1,\nu+1,\dots,w$. Furthermore in (1153) we have $\nu = 1,\dots,w$; $\varkappa = 1,\dots,a$.

Now choose

(1154) $$\sigma_{\nu+1} - \sigma_\nu \qquad (\nu = 1,\dots,w-1)$$

such that (1151) holds. Then the right-hand-side of (1153) is fixed. Now let $\sigma_1,\dots,\sigma_w$ simultaneously increase such that the differences (1154) remain unchanged. Then for sufficiently large $\sigma_1,\dots,\sigma_w$ the conditions (1152), (1153) are fulfilled. This proves (1150). For abbreviation set

(1155) $$P(s) = P(m,w,r,Z,s) .$$

If $\varphi \in \mathcal{Y}(w)$ then $\mathcal{W} = \varphi(\mathcal{W})$ hence

$$\varphi(\mathcal{V}) \cap \mathcal{W} = \varphi(\mathcal{V} \cap \mathcal{W}) \neq \emptyset .$$

Hence the set

(1156) $$i^* = \mathcal{J}(\mathcal{W} \cup \bigcup_{\varphi \in \mathcal{Y}(w)} \varphi(\mathcal{V}))$$

is connected and by theorems 104, 109, 110 the function $P(s)$ is holomorphic in $i^* + i\,\mathbb{R}^w$. The domain $\varphi(\mathcal{V})$ is described by the inequalities

(1157) $$\sigma_{\varphi(1)} + \sigma_{\varphi(2)} > d ,$$

(1158) $$\sigma_{\varphi(\nu+1)} - \sigma_{\varphi(\nu)} > d \qquad (\nu = 1,\dots,w-1) .$$

But since a fundamental domain of $\Gamma(w)$ in $\mathbb{R}^w$ is given by

(1159) $$\sigma_1 \leq \sigma_2 \leq \dots \leq \sigma_w ,$$

and since $\mathcal{J}$ denotes the convex hull, it is easily seen that the conditions (1158) may be dropped. Furthermore $\mathcal{V}^*$ contains a domain $\mathcal{V}$ described by

(1160) $$\sigma_\mu + \sigma_\nu > d \qquad (1 \leq \nu < \mu \leq w)$$

with suitable d. In $\mathcal{V} + i\,\mathbb{R}^w$ is P(s) holomorphic.

Now let τ be the substitution

(1161) $$\sigma_1 \to -\sigma_1,\ \sigma_\nu \to \sigma_\nu \qquad (\nu = 2,\dots,w) .$$

Because of $\tau(\mathcal{N}) = \mathcal{N}$ the set

(1162) $$\mathcal{G} = \mathcal{J}(\mathcal{V} \cup \tau(\mathcal{V}))$$

is connected and P(s) is holomorphic in $\mathcal{G} + i\,\mathbb{R}^w$. The domain $\mathcal{V}$ is given by

(1163) $$\sigma_\mu + \sigma_\nu > d \qquad (2 \leq \nu < \mu \leq w)$$

and

(1164) $$\sigma_\mu - \sigma_1 > d \qquad (\mu = 2,\dots,w).$$

The domain $\tau(\mathcal{V})$ is described by (1163) and

(1165) $$\sigma_\mu - \sigma_1 > d \qquad (\mu = 2,\dots,w) .$$

It is geometrically clear that the forming of the convex hull takes away the two conditions (1164), (1165). Hence $\mathcal{G}$ is simply described by (1163). Here σ_1 does no longer appear hence it may be arbitrary chosen in $\mathcal{G}$.
From the latter remark it is clear that

(1166) $$\mathcal{J}\Big(\bigcup_{\varphi \in \Gamma(n)} \varphi(\mathcal{G})\Big) = \mathbb{R}^w .$$

Hence $P(s)$ is holomorphic in $\mathbb{R}^w + i\,\mathbb{R}^w = \mathbb{C}^w$. It is invariant under the group $\Delta(w)$ which is generated by $\gamma(w)$ and τ. Theorem 111 is proved.

THEOREM 112: Let be $m = 1$ then the following functions are holomorphic in $\mathbb{C}^n$ and invariant under $\Delta(n)$:

(1167) $$P(1,n,0,Z,s) = \Big(\prod_{\nu=1}^{n} F(s_\nu + \tfrac{1}{2})\Big)\Big(\prod_{1\le\mu<\nu\le n}\{F(s_\nu+s_\mu+\tfrac{1}{2})F(s_\nu-s_\mu+\tfrac{1}{2})\}\Big)\mathcal{E}([1,n],0,Z,s),$$

(1168) $$P(1,n,r,Z,s) = \Big(\prod_{\nu=1}^{n} \xi(r,s_\nu + \tfrac{1}{2})\Big)\Big(\prod_{1\le\mu<\nu\le n}\{F(s_\nu+s_\mu+\tfrac{1}{2})F(s_\nu-s_\mu+\tfrac{1}{2})\}\Big)\mathcal{E}([1,n],r,Z,s) \quad (r > 0).$$

PROOF: Apply theorems 100 and 111.

REMARK 1: In axiom A the inequality (1092) has to be true. I do not know if this inequality follows from the inequalities given in Gričenko [12], page 596, theorem 3 and Maaß [34], page 236 formula (32).

REMARK 2: From theorems 85, 112 we see that all Eisensteinseries $\mathcal{E}(l,r,Z,s)$ for arbitrary l may be analytically continued by forming residues of Selberg's zetafunctions.

CHAPTER V. SIEGEL'S EISENSTEINSERIES

Siegel's Eisensteinseries are defined . It is shown that they may be derived from Selberg's Eisensteinseries (Representation with Selberg's zetafunction) by computing residues of Selberg's zetafunction as it was done in § 13. Hence the analytic continuation of Selberg's Eisensteinseries gives us analytic continuation of Siegel's Eisensteinseries. From the functional equations for Selberg's Eisensteinseries one gets a functional equation for Siegel's Eisensteinseries.

---

§ 19. SIEGEL'S EISENSTEINSERIES

Analytic continuation and the functional equation for Siegel's Eisensteinseries is obtained.

Like in (1054) define Siegel's Eisensteinseries by

$$(1169)\qquad E(n,r,Z,u) = \sum_{M \in \Gamma_n(n)\backslash\Gamma(n)} (\mathrm{Det}\ M\{Z\})^{-2r}(\mathrm{Det}\ Y_M)^{u-r} .$$

Obviously

$$(1170)\qquad E(n,r,M\langle Z\rangle,u) = (\mathrm{Det}\ M\{Z\})^{2r}\ E(n,r,Z,u) \quad (M \in \Gamma(n)) .$$

Set

$$(1171)\qquad \overset{(m)}{s}_{\nu}(u) = u - \frac{m+1}{4} - \frac{m}{2}(w-\nu) \qquad (\nu = 1,\dots,w) ,$$

$$(1172)\qquad \overset{(m)}{s}(u) = (\overset{(m)}{s}_{1},\dots,\overset{(m)}{s}_{w}) .$$

Then from (964), (1122) we deduce

$$(1173)\qquad |\overset{(m)}{s}(u)| = u - \frac{n+1}{4} .$$

An easy computation shows

$$(1174) \qquad \widetilde{\overset{(m)}{s}(u)} = \overset{(m)}{s}\left(\frac{n+1}{2} - u\right) .$$

THEOREM 113: Let axiom A be true and set

$$(1175) \qquad B(m,w,r,u) =$$

$$\left(\prod_{\nu=0}^{w-1} K\left(m,r,u - \frac{m\nu}{2}\right)\right)\left(\prod_{1 \leq \mu \leq \nu \leq w-1} C\left(m,2u - \frac{1+m(\mu+\nu)}{2}\right)\right) ,$$

$$(1176) \qquad R(m,w,r,Z,u) = B(m,w,r,u)E(n,r,Z,u) .$$

Then

$$(1177) \qquad R(m,w,r,Z,u) = \gamma(\operatorname{Det} Y)^{-2r}P(m,w,r,Z, \overset{(m)}{s}(u))$$

with some constant γ. The function $R(m,w,r,Z,u)$ is holomorphic for $u \in \mathbb{C}$ and it satisfies the functional equation

$$(1178) \qquad R(m,w,r,Z,u) = R\left(m,w,r,Z,\frac{n+1}{2} - u\right) .$$

PROOF: Let $\mathcal{Y}$ be a $\mathcal{Y}$-domain. For large enough $\operatorname{Re} u$ we have

$$(1179) \qquad \overset{(m)}{s}(u) \in \mathcal{Y} .$$

Hence theorems 87, 108 and formulas (1116), (1171), (1172), (1173), (1175), (1176) give (1177).

From theorem 111 it follows that $R(m,w,r,Z,u)$ is holomorphic for $u \in \mathbb{C}$. By theorem 111 the function $P(m,w,r,Z,s)$ is invariant under $\Delta(w)$. Hence

$$(1180) \qquad P(m,w,r,Z,s) = P(m,w,r,Z,\tilde{s}).$$

From (1174), (1180) we get (1178). Theorem 113 is proved.

THEOREM 114: The function

$$R(1,n,0,Z,u) = \left(\prod_{\varkappa=0}^{n-1} F\left(u - \frac{\varkappa}{2}\right)\right)\left(\prod_{1 \leq \nu \leq \mu \leq n-1} F\left(2u - \frac{\nu+\mu}{2}\right)\right)E(n,0,Z,u) \tag{1181}$$

is holomorphic for $u \in \mathbb{C}$ and satisfies the functional equation (1178).

PROOF: Apply theorem 100 and 113.

From (1181) one sees where $E(n,0,Z,u)$ can have poles. More about this may be found for arbitrary n in Christian [8] and Kalinin [19] and for $n = 2$ in Kaufhold [20] and Maaß [34].

THEOREM 115: Let $r > 0$. The function

$$R(1,n,r,Z,u) = \left(\prod_{\mu=0}^{n-1} \xi\left(r,u - \frac{\mu}{2}\right)\right)\left(\prod_{1 \leq \mu \leq \nu \leq n-1} F\left(2u - \frac{\nu+\mu}{2}\right)\right)E(n,r,Z,u) \tag{1182}$$

is holomorphic for $u \in \mathbb{C}$ and satisfies the functional equation (1178).

PROOF: Apply theorems 100 and 113.

§ 2o. POLES AND HECKE'S SUMMATION

It is shown that for certain small weights the analytic continuation of the Eisensteinseries is holomorphic. Hence for these weights one has Hecke summation.

From (151) we get

$$\xi(r,s) = \pi^{-s}\Gamma(s+r)\zeta(2s) \tag{1183}$$

where ζ is Riemann's zetafunction. Set

$$\xi(s) = \xi(0,s) = \pi^{-s}\Gamma(s)\zeta(2s) \, . \tag{1184}$$

Then the functional equation of the Γ-function gives us

$$\xi(r,s) = \left(\prod_{\nu=0}^{r-1} (s+\nu)\right)\xi(s) \, . \tag{1185}$$

Set

$$A(n,r,u) = \left(\prod_{\mu=0}^{n-1} \xi(r,u-\tfrac{\mu}{2})\right)\left(\prod_{1 \leq \mu \leq \nu \leq n-1} F(2u - \tfrac{\nu+\mu}{2})\right) \, . \tag{1186}$$

Then by theorem 115

$$R(1,n,r,Z,u) = A(n,r,u)E(n,r,Z,u) \tag{1187}$$

is holomorphic for $u \in \mathbb{C}$. From (1185), (1186) we deduce

$$A(n,r,u) = \left(\prod_{\mu=0}^{n-1}\prod_{\nu=0}^{r-1} (u - \tfrac{\mu}{2} + \nu)\right) \times \tag{1188}$$

$$\left(\prod_{\mu=1}^{n-1} \xi(u-\tfrac{\mu}{2})\right)\left(\prod_{1 \leq \mu \leq \nu \leq n-1} F(2u - \tfrac{\nu+\mu}{2})\right) \, .$$

THEOREM 116: The function

$$G(s) = s(\tfrac{1}{2} - s)\xi(s) = \frac{\tfrac{1}{2} - s}{1 - s} F(s) \tag{1189}$$

is holomorphic for $s \in \mathbb{C}$ and possesses no real zeros.

PROOF: See Landau [22], § 128.

THEOREM 117: It is

$$A(n,r,r) \neq 0 \quad (r = 1,2,[\tfrac{n-1}{2}],[\tfrac{n+1}{2}]) \, . \tag{1190}$$

For $n \geq 3$ the function $A(n,1,u)$ has a pole of first order at $u = 1$. For

(1191) $$3 \leq r \leq [\tfrac{n-3}{2}]$$

the function $A(n,r,u)$ has a zero of order

(1192) $$S(r) = \left\{ \begin{array}{ll} r - 2 & (3 \leq r < \frac{n+2}{4}) \\ & \\ [\frac{n-1}{2}] - r & (\frac{n+2}{4} \leq r \leq [\frac{n-3}{2}]) \end{array} \right\} .$$

PROOF: Express $\xi(s)$ and $F(s)$ by $G(s)$ and linear factors. Because of theorem 116 all real zeros and poles are given by the linear factors. Now the result follows by an elementary counting of poles and zeros. For another proof see Christian [8], [10].

THEOREM 118: It is

(1193) $$E(n,r,Z,r) \neq \infty \quad (r = 1,2,[\tfrac{n-1}{2}],[\tfrac{n+1}{2}]) ,$$

(1194) $$E(n,1,Z,1) = 0 \quad (n \geq 3) .$$

For (1191) the function $E(n,r,Z,u)$ has at $u = r$ at the most a pole of order $S(r)$.

PROOF: Apply theorems 115, 117.

THEOREM 119: Let $n \leq 8$. Then

(1195) $$E(n,r,Z,r) \neq \infty \qquad (r = 1,2,3,4) .$$

PROOF: Apply theorem 118.

Let $k \in \mathbb{Z}$ such that

(1196) $$E(n,r,Z,u) = (u-r)^{-k} D(n,r,Z,u)$$

and

$$D(n,r,Z,r) \not\equiv \infty . \tag{1197}$$

Then

$$D(n,r,Z,r) = \lim_{u \to r} \{(u-r)^k E(n,r,Z,u)\} . \tag{1198}$$

From (1170), (1198) we obtain

$$D(n,r,M\langle Z\rangle,r) = (\operatorname{Det} M\{Z\})^{2r} D(n,r,Z,r) \quad (M \in \Gamma(n)). \tag{1199}$$

If one can prove that $D(n,r,Z,r)$ is holomorphic in Z then for $n \geq 2$ this functions must be a modular form of weight $2r$.

It was proved by Weißauer [47] that

$$E(n,\tfrac{n+1}{2}, Z,\tfrac{n+1}{2}) \quad (n \equiv 1 \bmod 2) \tag{1200}$$

is holomorphic.

Theorem 118 may be also expressed that way, that the Eisensteinseries

$$\sum_{M \in \Gamma_n(n)\backslash\Gamma(n)} (\operatorname{Det} M\{Z\})^{-2r} \tag{1201}$$

possesses "Hecke summation" for

$$r = 1,2,[\tfrac{n-1}{2}],[\tfrac{n+1}{2}] . \tag{1202}$$

For the explanation of "Hecke summation" see Hecke [16], [17].

It is an important open question wether it is possible to take the poles away in the interval (1191). Of course for the function $K(m,r,u)$ in axiom A one may take $A(m,r,u)$ defined by (1186). Assuming that then (1091) holds one can make the whole computation with this $K(m,r,u)$. This leads again exactly to theorem 118. But if it is possible to find for a certain m a "better" $K(m,r,u)$ than our $A(m,r,u)$ then it may follow from theorem 113, that theorem 118 can be sharpend for all multiples $n = wm$. For $n = 2$ Gričenko [12] finds an $K(2,r,u)$, but this is identical with our $A(2,r,u)$ and cannot be used for sharpening theorem 118.

If it should be possible to find a $K(m,r,u)$ which helps to sharpen theorem 118 this will be only for the multiples of m. Hence the question arises if it is neccessary to assume (1051). If we start with an arbitrary l we may assume that there is a $K(l_\nu,r,u)$ for each $\nu = 1,\ldots,w$ such that the assertion of axiom A holds for each ν. I did not check what comes out if one gives up (1051). But this case becomes more difficult in so far as the functions $\lambda(l,Y,s)$, $\gamma_\nu(l,s)$ $(\nu = 1,2)$ (see (889), (890)) are not holomorphic but only meromorphic.

Let $q \in \mathbb{N}$ and χ a Dirichletcharacter mod q whith

(1203) $$\chi(-1) = (-1)^r .$$

Then Gričenko [12] considers for $n = 2$ the Eisensteinseries

(1204) $$\sum_{\substack{\cdots \\ c \equiv 0 \bmod q}} \chi(\mathrm{Det}\, D)(\mathrm{Det}(CZ + D))^r(\mathrm{Det}\, Y_M)^{\frac{s}{2}} .$$

He proves analytic continuation and it should be remarked (Gričenko does not mention this explicitely) that the function is holomorphic at $s = 0$ for $r = 1,2$, i. e., this Eisensteinseries has Hecke summation.

LITERATURE

[1] ANDRIANOV, A. N.: On zeta functions of systems of quadratic forms. Math. Notes Acad. Sci USSR 20, 571-674 (1974).

[2] ANDRIANOV, A. N.: Symmetric squares of zeta functions of Siegel modular forms of genus 2. Proc. Steklov Inst. Math. 142, 21-45 (1979).

[3] ANDRIANOV, A. N. and MALOLETKIN, G. H.: Behaviour of theta series of degree N under modular substitutions. Math. of the USSR Izvestija 9, 227-242 (1975).

[4] ARAKAWA, T.: Dirichlet series corresponding to Siegel's modular forms. Math. Ann. 238, 157-173 (1978).

[5] BRAUN, H.: Konvergenz verallgemeinerter Eisensteinscher Reihen. Math. Z. 44, 387-397 (1939).

[6] CHRISTIAN, U.: Über Hilbert-Siegelsche Modulformen und Poincarésche Reihen. Math. Ann. 148, 257-307 (1962).

[7] CHRISTIAN, U.: Siegelsche Modulfunktionen. 2. Edition, Göttingen 1980/81 (Lecture Notes).

[8] CHRISTIAN, U.: Bemerkungen zu einer Arbeit von B. Diehl. Abh. Math. Sem. Univ. Hamburg 52, 160-169,(1982).

[9] CHRISTIAN, U.: Eisenstein series for congruence subgroups of GL(n, ℤ). Amer. J. Math.

[10] CHRISTIAN, U.: On the analytic continuation of Eisenstein series for Siegel's modular group of degree n. Monatshefte f. Math.

[11] DIEHL, B.: Die analytische Fortsetzung der Eisensteinreihe zur Siegelschen Modulgruppe. J. reine angew. Math. 317, 40-73 (1980).

[12] GRIČENKO, V. A.: Analytic continuation of symmetric squares. Math. of the USSR Sbornik 35, 593-614 (1979).

[13] GUNDLACH, K.-B.: Dirichletsche Reihen zur Hilbert'schen Modulgruppe. Math. Ann. 135, 294-314 (1958).

[14] HARISH-CHANDRA: Automorphic forms on semisimple Lie groups. Springer Lecture Notes in Mathematics 62 (1968).

[15] HASSE, H.: Vorlesungen über Zahlentheorie. Springer Verlag, Berlin, Göttingen, Heidelberg 1950.

[16] HECKE, E.: Analytische Funktionen und algebraische Zahlen. Zweiter Teil. Ges. Abh. 381-404 and Abh. Math. Sem. Univ. Hamburg 3, 213-236 (1924).

[17] HECKE, E.: Theorie der Eisensteinschen Reihen höherer Stufe und ihre Anwendung auf Funktionentheorie und Arithmetik. Ges. Abh. 461-486 and Abh. Math. Sem. Univ. Hamburg 5, 199-224 (1927).

[18] HÖRMANDER, L.: An introduction to complex analysis in several variables. D. van Nostrand Company. Princeton, N.J., Toronto, London.

[19] KALININ, V. L.: Eisenstein series on the symplectic group. Math. of the USSR Sbornik 32, 449-476 (1977).

[20] KAUFHOLD, G.: Dirichlet'sche Reihe mit Funktionalgleichung in der Theorie der Modulfunktion 2. Grades. Math. Ann. 137, 454-476 (1959).

[21] KUBOTA, T.: Elementary theory of Eisenstein series. Kodansha, Tokyo and John Wiley, New York, London, Sydney, Toronto 1973.

[22] LANDAU, E.: Handbuch der Lehre von der Verteilung der Primzahlen. Chelsea Publ. Comp., New York 1974.

[23] LANGLANDS, R. P.: Eisenstein series, Proc. Sympos. in pure math. 9, 235-252 (1966).

[24] LANGLANDS, R. P.: On the functional equations satisfied by Eisenstein series. Springer Lecture Notes in Mathematics 544.

[25] MAASS, H.: Automorphe Funktionen und indefinite quadratische Formen. Sitzungsber. Heidelberger Akad. Wiss. 1949, 1-42.

[26] MAASS, H.: Über eine neue Art von nichtanalytischen automorphen Funktionen und die Bestimmung Dirichlet'scher Reihen durch Funktionalgleichungen. Math. Ann. 121, 141-183 (1949).

[27] MAASS, H.: Modulformen zweiten Grades und Dirichletreihen. Math. Ann. 122, 90-108 (1950).

[28] MAASS, H.: Die Differentialgleichungen in der Theorie der elliptischen Modulfunktionen. Math. Ann. 125, 235-263 (1953).

[29] MAASS, H.: Die Differentialgleichungen in der Theorie der Siegelschen Modulfunktionen. Math. Ann. 126, 44-68 (1953).

[30] MAASS, H.: Lectures on Siegel's modular functions. Bombay 1954 - 55.

[31] MAASS, H.: Lectures on modular functions of one complex variable. Bombay 1964.

[32] MAASS, H.: Some remarks on Selberg's zeta functions. Proc. intern. conf. on several complex variables. U. of Maryland, College Park, Md. 1970, 122-131.

[33] MAASS, H.: Siegel's modular forms and Dirichlet series. Springer Lecture Notes in Mathematics 216.

[34] MAASS, H.: Dirichlet'sche Reihen und Modulformen zweiten Grades. Acta Arith. 24, 225-238 (1973).

[35] NEUNHÖFFER, H.: Über die analytische Fortsetzung von Poincaréreihen. Sitzungsberichte Heidelberger Akad. Wiss. 1973, 33-62.

[36] RAGHAVAN, S.: On Eisenstein series of degree 3. J. Indian math. Soc. 39, 103-120 (1975).

[37] ROELCKE, W.: Das Eigenwertproblem der automorphen Formen in der hyperbolischen Ebene I, and II. Math. Ann. 167, 292-337 (1966) and 168, 261-324 (1967).

[38] SATO, F.: Zeta functions in several variables assosiated with prehomogeneous vector spaces. Proc. Japan Acad. 57, part I: 74-79, part II: 126-127, part III: 191-193 (1981) and
I: Tôhoku Math. J. 34, 437-483 (1982)
II: Tôhoku Math. J. 35, 77-99 (1983)
III: Annals of Math. 116, 117-212 (1982).

[39] SELBERG, A.: Harmonic analysis 2. part. Lecture Notes Göttingen 1954.

[40] SELBERG, A.: Harmonic analysis and discontinuous groups in weakly symmetric Riemannian spaces with applications to Dirichlet series. J. Indian Math. Soc. 20, 47-87 (1956).

[41] SELBERG, A.: A new type of zeta functions connected with quadratic forms. Report of the Institute in the theory of numbers, Boulder, Colorado 1959, 207-210.

[42] SELBERG, A.: Discontinuous groups and harmonic analysis. Proc. Intern. Congr. Math. 1967, 177-189.

[43] SHIMURA, G.: Confluent hypergeometric functions on the tube domains. Math. Ann. 260, 269-302 (1982).

[44] SIEGEL, C. L.: Lectures on advanced analytic number theory. Bombay 1961.

[45] TERRAS, A.: A generalization of Epstein's zeta function. Nagoya Math. J. 42, 173-188 (1971).

[46] TERRAS, A.: Functional equations of generalized Epstein zeta functions in several complex variables. Nagoya Math. J. 44, 89-95 (1971).

[47] WEISSAUER, R.: Eine Verallgemeinerung eines Satzes von Raghavan. To appear.

LIST OF SYMBOLS

Lower case latin letters

Capital latin letters

L.c. Greek letters

Capital Greek letters

Other symbols

INDEX